THIS IS WILDFIRE

HOW TO PROTECT YOURSELF, YOUR HOME, AND YOUR COMMUNITY IN THE AGE OF HEAT

NICK MOTT and **JUSTIN ANGLE**
Illustrations by Jessy Stevenson

BLOOMSBURY PUBLISHING
NEW YORK · LONDON · OXFORD · NEW DELHI · SYDNEY

BLOOMSBURY PUBLISHING
Bloomsbury Publishing Inc.
1385 Broadway, New York, NY
10018, USA

BLOOMSBURY,
BLOOMSBURY PUBLISHING,
and the Diana logo are trademarks of
Bloomsbury Publishing Plc

First published in the United States 2023

While the authors have consulted
experts in fire preparedness, nothing
makes a home or community entirely
fireproof. This book is a starting point
for creating resilience to wildfire, but
is not comprehensive. Living in a
wildfire-prone area is a calculated risk
and warrants continual thought and
assessment.

ISBN: HB: 978-1-63973-079-7;
eBook: 978-1-63973-080-3

Library of Congress Cataloging-in-
Publication Data is available

2 4 6 8 10 9 7 5 3 1

Designed and typeset by Sara Stemen
Printed and bound in the U.S.A.

To find out more about our authors and
books visit www.bloomsbury.com and
sign up for our newsletters.

Bloomsbury books may be purchased
for business or promotional use.
For information on bulk purchases
please contact Macmillan Corporate
and Premium Sales Department at
specialmarkets@macmillan.com.

This book is dedicated to the people trying to change how we live with fire.

From tribal members to firefighters to government employees to scientists to community members, people all over the country have started a movement to reestablish a healthy coexistence with flames. Those who champion that movement are carrying a torch that grows ever brighter.

CONTENTS

MANAGING

INTRODUCTION

In late August 2020, lightning struck a tree near Bozeman, Montana. It's a pretty ordinary occurrence in late summer across the West. For days, the tree just smoldered, too small to attract any notice. Then a gusty heatwave hit the area. Soon the fire had spread to around ten acres. By the next morning, it had grown about forty times bigger, burning over four hundred acres. Then the fire exploded. That evening, it ran across the landscape. Officials later estimated it to be about seven thousand acres.

The blaze was right outside the city, on a popular hiking trail adorned with a massive, 250-foot white "M." From most of town, residents could see the smoke plume and hear the sirens and the buzz of planes and helicopters soaring overhead. It made headlines across the state.

It was a small fire, at least by the standards of the last decade or so—about eight thousand acres in the end. But it burned hot and fast and shocked the people living nearby. The fire was just a few miles from a popular ski resort. Some folks had minutes to evacuate. Others prepared a day or hours earlier. People lost pets and memories. They drove to a Red Cross shelter in town with no idea of the fate of their homes. All told, it destroyed about thirty houses.

As the fire made a mad dash across the landscape, firefighters on the ground couldn't get out of its way. Some members of one fire crew had to deploy their "shelters," shiny silver blankets meant to protect them from unfathomable heat when no other options exist. One crew member had to share a shelter with a companion.

An October storm brought precipitation and cool weather to the area. Over a month after it began, the fire was finally under control.

A year later, a different story of the Bridger Foothills fire emerged. Driving the winding canyon road nearby, FOR SALE signs sprouted, almost like flowers. Tyvek siding grew like moss on homes, the telltale sign of immediate rebuilding.

Hiking up that popular trail on which the fire began, green burst through the black, matchstick trees. The soil was loose and smelled of ash. Wildflowers brought color to the earth—glowing, golden balsamroot and purple lupine. Hikers who venture off-trail might find precious morels. People all over the West forage for the tasty mushrooms the year after burns, their hands and knees blackened by ash. If you stopped and listened, you might've heard the *pock-pock-pock* of a black-backed woodpecker, a species that depends on new burns to survive.

The landscape was coming back to life.

There are at least two stories of every major wildfire (and sometimes many more). One story is about the destruction wrought by the flames: the acres burned, what it meant to human life and property. The other story is about the fire's impact on the ecosystem. The two stories are often at odds: fire is at once destructive and life-giving. It's something we seek to suppress and control, yet it's a natural force we need more of. This book is about reconciling those two views of flames and finding a way to live—happily, healthily, and sustainably—with our fiery future.

———————

Wildfires are getting bigger and more destructive than at any time in living memory. Flames—started by campers, cars, power lines, lightning, lighters, and more—engulfed more than ten million acres in 2020. That's a land mass fifty times larger than New York City and second (barely) only to 2015 in total acres burnt since reliable statistics began. Federal agencies started tracking wildfire acreage nationally in the early 1980s. By 1990, the biggest fire year on the books was '88—at around five million acres scorched that year. By the new millennium, there were three more record fire years on the books—all over five million acres. The acreage burnt in 2000 was staggering: more than seven million. Then, fires kept getting bigger. Twelve of the next twenty-one years broke the seven-million-acre mark. New records were established. Three years since 2010 broke ten million acres. The unthinkable is now the norm.

While that number is striking, the total area burnt isn't always the best indicator of the severity of a fire season, or even a single fire. Size doesn't necessarily equal destructiveness. Even a relatively small fire can bring horrendous damage to lives, livelihoods, and homes. Take, for example, 2018: 8,767,492 acres burnt across the country. That's a lot—but these days, it's not record-breaking. However, that same year, fires destroyed more than twenty-five thousand structures; 70 percent of them were homes. California got the brunt of it. The Camp Fire, a 150,000-acre megafire in the northern part of the state, was alone responsible for searing nearly fourteen thousand homes, losses worth more than $16 billion. Eighty-five people died. In fact, nineteen of the twenty most destructive fires—as defined by the number of buildings destroyed—in California history have occurred since 1999, and from 2005 to 2020 wildfires incinerated nearly ninety thousand structures. A recent report estimated that the cost per year of wildfire in the U.S.—for everything it destroys, and for putting it out—averages between about $70 billion and $350 billion. That's between ten and fifty times the budget for the entire U.S. Forest Service in 2021.

Nearly year-round, you can find headlines about wildfires ripping through towns, upending lives, sending smoke across the country. They're portrayed as terrifying, devastating, almost evil. The losses they produce are, by nearly any measure, tragic. People lose their lives to fires. They displace entire communities, kill pets, and change landscapes that people have lived their entire lives with. There are other impacts, too, that aren't quite so quantifiable: the anxiety of residents packing their bags, wondering when they need to leave their lives behind and hit the road. The terror of not knowing if loved ones, pets, homes, and livestock are okay. Burning lungs and coughing from days, weeks, or months of gulping down smoke. The consequences of these now normal megafires don't just affect people living in areas susceptible to fire, or even just people in the West. In recent years, wildfire smoke from the American West has made its way across the country, bringing an eerie and dystopian feel to places as distant as New York City. Fires frequently get so big they generate their own weather systems. In 2020, wildfire shut down I-70, a busy interstate in the heart of Colorado, for two weeks, disrupting national supply chains. Other fires have caused chaos for important freight routes on railroads. The Cerro Grande fire in 2000 threatened Los Alamos National Laboratory, a nuclear weapons facility in New Mexico. Every fire season, numerous weddings, camping trips, and other outdoor excursions get postponed, often devastating rural economies reliant on tourism. And while the population of wildfire-prone areas is booming, some people with enough resources are moving away, fed up after years of evacuations and smoke.

So why are we experiencing such devastating, enormous wildfires? It's not that we're seeing drastically *more fires* than ever before—the data show that the number of fire starts each year varies widely, but isn't changing significantly in one way or the other. However, more and more of these fires are ripping out of control into enormous, destructive conflagrations.

The dynamics of our fiery future are complicated. But it's helpful to think about three main drivers of this future, all human caused. First is the way we've treated fire for more than a century. Starting in the early days of forestry in the U.S., wildfire became an evil that federal and state agencies sought to snuff out with military-style efficiency. For decades, our national firefighting apparatus excelled at that mission. At the same time that fires were extinguished, that absence of fire shaped our landscapes. Trees encroached on areas typically dominated by shrubs and bushes. Tiny saplings popped up in the midst of trees that had been around for centuries, clogging up our forests. This all served as fuel for fire. Overgrown and dense, forests were ready to go up in smoke with a ferocity that was new on the landscape.

The second driver is where we've developed. Seeking solace in wild places or priced out of city centers, homeowners and renters have moved to areas ready to burn. In the world of wildfire, these areas are called the wildland urban interface, or WUI (pronounced "woo-ee"), where homes intermingle with flammable stuff like trees and underbrush. More homes and properties and things people are attached to than ever before reside in areas ripe for fire.

Climate change is the third culprit for putting us in this modern era of megafires. Skyrocketing rates of greenhouse gas emissions over the last three centuries or so have created global surface temperatures more than 1 degree Celsius warmer than before industrialization. That heat, along with climate-influenced drought, is generating conditions that make matchsticks of our forests.

Humans have made wildfires more destructive, but wildfires have been around longer than people. Trees, plants, and animals depend on burnt landscapes. Indigenous people used fire for millennia for hunting, harvesting, and ceremony before white settlers sought to put it out—and some of those traditions continue and are being revived today. Not all fire should be extinguished. If fire destroys, it also heals.

––––––––––

This book is organized into four sections:

"Burning" focuses on how we got here. It explores fire's impact on humanity, and in turn our influence on fire. It explores the drivers of today's wildfire crisis. Part of that story involves the Big Burn, which ignited a century of fire suppression policy. But it also entails the evolution of science to understand fire as sometimes beneficial, even essential.

"Fighting" focuses on the technology, techniques, and human effort we've dedicated to controlling flame. Tens of thousands of firefighters across the country fly, hike, parachute, and rappel into the flames. Firefighters are suffering from increasing rates of injury and burnout, as well as low wages. What used to be fire seasons are now "years." In the fire business, a "career fire" refers to a burn so big you'd see it one time in your decades-long tenure fighting flames. Now, many firefighters say they've been on several career fires over the last decade alone. How do firefighting professionals actually work on the ground? And what does that work mean for their well-being?

"Managing" focuses on trees and forests. What constitutes a "healthy" forest? Forest thinning projects are touted by some as a crucial component of wildfire resilience, yet environmentalists claim that they're often little more than logging by another name. How can we manage forests and fires in a way that's good for both people and the planet?

"Adapting" focuses on solutions. Fire seems big and unstoppable. Indeed, it's here to stay. Solutions aren't only in the hands of policymakers and foresters, and our future doesn't have to be all smoke and flame. People living in fire-prone landscapes can make decisions that will affect how they and their communities survive and thrive in the future.

Animating these sections is the idea that we all need to reformulate our relationship with fire.

One day this summer, a lightning bolt will strike an old ponderosa pine. Or a gust of wind will send a power line to the ground. Or a car will light up some vegetation on an overgrown double-track road. A train will send off a spark. The burning entrails of a firework will rain down on a juniper. Someone will take a lighter to a blade of grass.

There's no way to eliminate fire. But we can mitigate its severity and how it affects us. Doing that requires thinking long term. It requires changing ourselves and our homes and engaging with the world around us and our neighbors in ways we may not have considered before.

Fire season is at least eighty days longer now than it was thirty years ago. By the end of the century, experts anticipate extreme fire events to rise by 50 percent across the world. That means flames will burn bigger areas during more parts of the year, from temperate grassland to tropical savanna to tundra—in Australia, Greece, the Amazon, Indonesia, the Arctic, and beyond.

We've mobilized an army to fight a never-ending war on fire, yet we build deeper into areas ready to go up in flames. And we keep ramping up our burning of fossil fuels, contributing to the warming of our planet and making conditions ripe for bigger, more extreme blazes.

We're at a precarious moment for wildfire in the U.S. We can keep thinking about these conflagrations the same way, or we can change course. The question driving this book is: How can we recognize both truths of wildfire? It threatens and destroys so much of what we love: our homes, livelihoods, mental state, our very ability to breathe. But it's also a process that's completely natural. It rejuvenates forests and ecosystems. Fire has burned across the earth longer than our species has walked the planet. How do we live in a world with fire? How do we manage it? And how do we manage ourselves?

BURNING

Humans have had a relationship with fire since our evolutionary beginnings. Fueled by the Industrial Revolution, that love affair has heated up in the last few centuries. Fire has shaped nearly every aspect of our culture and society and played an enormous role in our politics. To understand how we can exist in a fiery world, we need to understand how we got here in the first place.

FIRE AND PEOPLE: A SHORT HISTORY

"How lucky that Earth has fire."
—RICHARD WRANGHAM

In terms of planetary existence, human history is a spark and wildfire a steady flame.

Fossilized charcoal suggests that wildfire emerged more than 400 million years ago. That's 399,700,000 years before humans came about. It was even before dinosaurs. At the time, a previously turbulent and erratic global climate had stabilized. Glaciers had melted and sea levels had risen. Only relatively simple life-forms could survive on earth. Coral reefs began to grow. Mollusks, trilobites, and enormous sea scorpions swam the ocean. The first beings that resembled spiders and centipedes crawled on land. Plants took root. Moss grew alongside water.

With the emergence of plants, two ingredients had taken shape to allow fire. Plants served as fuel, and there was enough oxygen in the atmosphere to fan the flames. Around 380 million years ago, early forms of trees came into existence. Soon after, coastal forests sprang into being. For millions of years, the prevalence of fire ebbed and flowed based on the presence of oxygen in the air. In that time, fire came sporadically, mostly through lightning strikes, and at times through volcanic activity, sparks from rockfall, and even, possibly, meteor impacts. Plants never existed without fire. And fire never existed without plants. In this time, fires burned and went out without interference from any living being. There's no evidence that any creature made use of fire.

Then, a little less than two million years ago, big-brained bipeds took an evolutionary leap. *Homo habilis*, the first species in the genus *Homo*, gave way to *Homo erectus*. This new species was more similar to modern-day humans than anything that came before. It stood upright and walked and ran in much the same way we do. For decades, scholars theorized about what triggered these drastic changes that sent us into modernity. Since the 1950s, the most popular anthropological answer was: *meat-eating*. The rise of eating raw meat likely fostered bigger brains, more cooperation, long-distance travel, and more.

But the Harvard anthropologist Richard Wrangham argues that that explanation alone is insufficient. *Homo erectus*, he says, had small teeth and small mouths—much like we do today—poorly adapted to eating raw meat. There had to be *something else*. What was it?

He argues that one thing alone could have propelled us into humanity: cooking our food, made possible by our control of fire. "We humans are the cooking apes, the creatures of the flame," he writes. Fire wouldn't have been completely foreign to these early hominids. Wildfires had been burning for millions of years. In addition to naturally started fires, some flames could have come from the strikes of flint during toolmaking. At some point, Wrangham argues, early hominids figured out how to contain and control those flames. When they did, a change was set in motion that affected these creatures' brains, anatomy, and social lives.

Cooking food to eat is something we take for granted today. Its benefits seem clear. But for anthropologists, the real benefit of cooking food was an evolutionary mystery for decades.

The established science of the time didn't have much to say on the topic. Scientists believed that cooked food had the same caloric value as raw food. But Wrangham devised a series of experiments to test that theory. He began with mice to see whether eating cooked food would create measurable differences in how much the animals grew.

The results were striking. Mice eating cooked food gained significantly more weight compared with mice eating raw food. He controlled for myriad other variables—amount of exercise, amount of food, all the alternative explanations he could think of. The results were clear: at least in mice, cooked food produced larger animals. The same holds true for domestic animals like calves and piglets. Apes, too, prefer cooked food over raw food when given the choice.

Based on that critical finding and others, Wrangham hypothesizes that cooking makes food easier to digest. The body doesn't have to use as much energy to process cooked food as it does raw food. It also means that the eater simply spends a lot less time doing what Wrangham calls "the physiological work" of eating food. He points to the novice wildlife photographer. Spending days in the jungle in pursuit of glorious chimpanzee photos, this photographer might be frustrated by the sheer amount of time chimps spend eating. Chimps in Tanzania's Gombe Stream National Park, for example, spend about six hours a day chewing. It takes a lot of chewing to soften up raw food.

But cooking makes food easier to chew, swallow, and digest. Wrangham says humans spend between a fifth and a tenth of the time chewing as do great apes. Early hominids would have been able to use that free time to do *other stuff*—hunting, gathering, socializing—and getting better at all of it.

Wrangham also hypothesizes that fire could have been crucial in helping early hominids leave the trees and spend time on the ground. It could have helped illuminate camps at night, revealing predators and other threats. It would have helped early humans keep warm on chilly evenings. That temperature regulation could have allowed us to shed our apelike hair and gain the appearance we have today.

Cooked food, Wrangham argues, likely even played into our developing intelligence. After all, it takes a tremendous amount of energy to run a big brain. Our brains use up about 20 percent of our resting energy but make up only about 2 percent of our body weight. Most other mammals use about half as much energy on their brains as humans. But early hominids developed smaller guts. Cooked food didn't take as much work to digest as raw food—and, Wrangham says, the energy saved could be used to fuel a larger brain.

If Wrangham is right, fire shaped our organs, muscles, skeleton, teeth, and brains. Fire made us what we are.

SHAPING FIRE

Look around your home, or wherever you're reading this. Try to count everything using fire. There are the obvious things: your grill, your fireplace, your stove. Even your car requires a spark and internal combustion. The rest is a little more hidden: the electricity that keeps your lights on, the blue flame of the pilot light in your furnace.

Early humans harnessed flames to cook. That would've meant burns akin to campfires today. At first, that might have meant cooking over naturally occurring fires. Eventually, we figured out how to light and control flames for ourselves. Over time, that control of open flame grew more efficient. We honed our cooking methods, and we went farther. People set fires to clear earth and fertilize soil.

The ways we use fire came to define how scholars categorize our human development. The Bronze Age began about five thousand years ago, when humans could produce heat high enough to forge copper and tin alloys. A couple thousand years down the road came the Iron Age. Civilizations figured out how to construct specialized furnaces that could smelt iron ore to create tools and weapons that would change the world.

Flame changed the world once more in the 1700s, and we're still seeing the effects of that change today. In the late eighteenth century, the steam engine brought power to England and ushered in the Industrial Revolution. The new technology upped the speed and efficiency with which goods could be produced, and it set in motion a chain of events that would lead to the world as we know it today. Societies urbanized. Before long, more powerful and efficient versions of the steam engine were in development. We learned to create electricity. In the mid-1800s came the first internal combustion engine, another innovation in how we contain and manipulate heat and flame that laid the groundwork for the automobile.

Fire has been part of humanity's history from its evolutionary beginning, Wrangham argues. Some carnivorous birds have been observed carrying flaming sticks to start fires and scare up prey. And chimpanzees in Senegal can predict the spread of wildfires in the savanna and react to it. But no other being on the planet creates, manipulates, and controls fire the way humans do.

Imagine a campfire or fireplace. It's easy to lose yourself in the flames, to stare, engrossed in the orange blue licking the air. Even in a group, fire makes silence and contemplation comfortable. The fire historian Stephen Pyne, an emeritus professor at Arizona State University and former firefighter, says that sensation is deeply human. The word *focus*, for example, comes from a Latin term for "fireplace" or "hearth." Fire has made its way into our metaphors: we experience the fires of love, romance, and passion. We fan the flames. We blaze a trail. We fight fire with fire. "Other animals knock over trees and dig holes in the ground," he says. "We do fire."

CLIMATE AND FIRE

In less than three hundred years, the United States alone has emitted more than four hundred billion tons of carbon dioxide. Preindustrial levels of carbon dioxide in the atmosphere were around 278 parts per million. By 2022, that number had nearly doubled—at over 414 ppm. The last time in global history that atmospheric CO_2 levels were that high, humans didn't exist. Sea levels were seventy-five feet higher than they are today. Much of the Arctic was covered in forest. The culprit? Burning.

The other side of the mastery of fire that made us what we are had major, unintended consequences. The story of humans and fire is also the story of industrialization. The burning that powers our world also threatens our very existence. One could easily trace the relationship of humans and fire as one of ongoing pursuit of control. But now, that control has escaped us. In a very tangible sense, as we see more acres burnt and more structures incinerated, we've lost the control we once exercised over wildfire itself. In a more subtle way, too, the changing climate is completely bound up with humanity's addiction to burning.

"It's really about our relationship with combustion," says Jennifer Balch, an associate professor of geography at the University of Colorado Boulder. She says it all links back to climate change. Dry fuels during fire season are due to the greenhouse gases we've pumped into the atmosphere over the last couple hundred years. "And then I see a fire start because a power line goes down just like, wow, both of those elements are related to our energy addiction."

Downed power lines in California have caused more than fifteen hundred fires over the last seven years, including the Camp Fire, the deadliest in the state's history. They're responsible for six of the twenty most destructive fires in the state since 2015. In Oregon in 2020, power lines started at least thirteen fires during a single windstorm. Climate change is poised to heighten that risk even more. Scholars say extreme weather associated with climate change can create conditions ripe for power line failure—like those gale-force winds in Oregon.

Eight of the ten warmest years on record in the U.S. have occurred since 1998. During this time, the country has seen its most extreme fire behavior on record. Data on California's wildfires, for example, go back to 1932. Nine out of ten of the largest fires in the state's history have occurred in the last decade. From at least recent data, the correlation between a warming climate and wildfire seems strong. But our data on wildfires are relatively short term. We can rigorously look only a few decades into the past at how much land burned each year. Is there a way to get a longer-term perspective on the relationship between earth's fire and temperature? Looking back at the deep past is Cathy Whitlock's wheelhouse. She's a paleoecologist at Montana State University. That means she studies really old stuff to understand how our climate used to be. Then she uses the data she unearths to understand what our climate is experiencing now.

She and her team take cores of mud from lake bottoms. The deeper the core, the older the mud. Over tens, and in some cases, hundreds of thousands of years, artifacts of the ecosystem fall into the lake and settle into that mud. She's dug cores from Argentina to Tasmania to Yellowstone National Park.

Each layer of the mud core is a time capsule: a glimpse of what the ecosystem was like long, long ago.

When the core gets to her lab in Montana, the analysis begins. Whitlock is interested in what's *inside* that mud. In particular, pollen traces show what kinds of plants were around in a given time period—as in pine, sagebrush, or tundra plants. As she analyzes layer after layer, she says, "What you see is the composition of the pollen assemblage changes through time. And that's telling you how the vegetation is changing, it's changing in response to climate. Pollen is the best way of reconstructing past climate, going back over those timescales."

Along with pollen, her team also searches for traces of charcoal. When wildfires burn on a landscape, bits of ash and coal fall from the air—just like pollen—and descend to the depths of those lakes. While she reconstructs the history of vegetation through pollen, she can also reconstruct the history of fires.

From all the data about the climate and wildfires tens of thousands of years ago, a distinct trend emerges: "When we look at the past, we always see that there are more fires when the climate is warmer," Whitlock says.

The science behind how a changing climate leads to our worsening wildfire crisis is manifold. One of the main factors in the relationship between fire and climate is a concept scientists call "vapor pressure deficit." Technically speaking, VPD refers to the difference between how much moisture is in the air and how much moisture the air *could* hold if it was fully saturated. Functionally, that number details how easy it is for the atmosphere to slurp moisture out of the soil. If the air is a blow dryer, that number tells you how powerful its drying action is. The higher the temperature and the lower the humidity, the more moisture the atmosphere can sponge out of the earth, drying out all the sticks and leaves and shrubs and trees—in fire lingo, the *fuel*.

Here's an example. California reached its highest vapor pressure deficit in at least forty years in August 2020. That same month, according to Cal Fire, California's state firefighting and forestry agency, there were fifty-nine fire starts. Those included the August Complex, which would become the state's largest fire on record and burn over a million acres. A separate fire, the North Complex Fire, scorched an area about the size of New York City in a single night.

Unless more moisture makes its way into the earth's atmosphere, VPD inherently rises with temperatures. In other words, as the earth warms, VPD goes up. One 2017 study predicts a "continental-scale drying of the United States atmosphere," and specifically a 51 percent increase in vapor pressure deficit in summertime by 2100.

As overall temperatures are rising, in many areas the VPD difference between night and day is narrowing too. The consequences of that could be enormous. When the sun goes down, wildfires ordinarily calm down. As night falls, temperatures lower, less moisture evaporates from the soil, and fuels grow slightly moister. That can create a natural reprieve from the spread of flames. Wildland firefighters often take advantage of that cycle and use evening hours to get ahead of the burn. Studies of nighttime wildfire behavior are in their relative infancy. But one 2022 publication shows that the daily minimum of VPD went up by

25 percent from 1979 to 2020. The western U.S., the study found, was a hotspot: the number of flammable nights went up by 45 percent from 1979 to 2020—nearly twice what the globe as a whole experienced—and the intensity of night fires increased by 28 percent, about four times as much as the rest of the planet. The overall trend, the authors say, indicates that we should expect wildfires to be "more intense, longer-lasting, and larger." The authors conclude that climate change is causing the failure of the "night brakes" that naturally dampen the spread of fire.

In addition to vapor pressure deficit, seasonal changes brought on by a warming climate also affect wildfires. In wintertime, snow starts to pile up on mountaintops, covering up fuels and providing a reservoir of moisture for the hot summer ahead. But as the climate warms, snowpack decreases in the high mountains. In spring, what used to be spring snow falls as rain, sheeting off snowpack earlier and faster. Studies show that snowpack is melting away one to four weeks earlier than fifty years ago. Earlier runoff can also portend drought later in the summer. It also means those high-elevation forests that would ordinarily be covered in snow are drier and more vulnerable to fire.

Summers are getting hotter and drier. The West as a whole is experiencing its worst drought in twelve hundred years. "By the end of the summer, in the last few years, the trees are so dried out that they're almost like standing matchsticks," Whitlock says. "They're just ready to go."

Even in the fall, the changing climate is driving more wildfires. In California, the Santa Ana winds exacerbate fires just about every fall. Those winds are dry gusts that blow westward from the arid desert toward low-pressure systems on the Southern California coast. Recent climate models suggest that those winds could actually become less frequent in the fall. Yet the winds will still come in winter—and as patterns of rainfall change with the warming climate, scientists hypothesize that could create an even longer winter fire season in California.

The numbers bear this out. Fire season is nearly eighty days longer than it was back in 1970. Scientists expect that to grow even more—by another three weeks in the next thirty years. Whitlock says climate change isn't just influencing the frequency and severity of burns. A big question now is what happens *after* a fire, when things start to grow again. "The climate is changing. So the conditions where the previous forest was established don't exist anymore," Whitlock says. "And we're already seeing that those species don't necessarily return, but new things come in."

Especially at low elevations, plants like juniper and sage start to appear, rather than ponderosa pine. There's grassland where there once was forest. So a fire sweeps in, and since the temperatures are rising, that one burst of flame can change the whole ecosystem. One 2019 study suggests that by the middle of the century in the Rocky Mountain West, more than 6 percent of all forests are at risk of never coming back if an extreme fire comes through due to climate change. If you look a little farther down on the map, to the Southwest, that number jumps to 30 percent.

"My perspective as a paleoecologist is that we've seen really big climate changes, but it takes place over centuries to thousands of years," Whitlock says. "But what's happening now is super fast. And the whole ecosystem really is, I think, in disequilibrium as a result."

One 2021 study found that subalpine ponderosa forests in the Rocky Mountains are burning more than at any point in the last two thousand years. Subalpine forests in the Rockies lie between nine thousand and eleven thousand feet and are often dominated by fir and spruce.

Phil Higuera, a professor of ecosystem and conservation sciences at the University of Montana and lead author on that study, says the work emerged from 2020's horrendous fire season. That year, more than ten million acres burned across the U.S.; more than four million of them in California alone. That extreme fire behavior prompted his team to ask the question: Is what we're seeing unprecedented in the past?

"For most of my career, when we look at the past—in a comforting way, we see that, oh, these things that are unusual in the human timescale, they've happened before," Higuera says. "But this paper was different."

Higuera says that the messaging is complicated—because lower-elevation ponderosa pine forests are burning less than ever before, largely due to a century-old policy that encourages fire suppression. "Some areas are burning more than they ever have, others aren't burning enough," Higuera says. Human activities—whether in the form of development, fire suppression, or climate change—are changing fire regimes in nearly all ecosystems. "Warming the climate is loading the dice," Higuera says.

That problem comes back to our addiction to burning and therefore carbon and other emissions we release into the atmosphere. Forests are carbon *sinks*. They trap carbon and keep it in the ground. In 2021, forests in the U.S. stored about 61 billion metric tons of carbon. But when wildfires char a forest, that carbon pulses up into the air. In that same year, wildfires belched out more than 1.7 billion tons of carbon dioxide into the atmosphere. That's about twice as much as the entire country of Germany. Our carbon sinks are turning into sources of emissions.

When massive fires burn, smoke and ash can enter the atmosphere and travel for thousands of miles. Scientists, for example, have tracked smoke from Canadian wildfires all the way to the Greenland Ice Sheet. Projections show that melt from the 660,000-square-mile ice cap could fuel more than twenty feet of sea level rise. As fire soot lands atop the ice, it blackens the white landscape. Instead of reflecting heat back into the atmosphere, it absorbs it. Although there's not yet a clear long-term relationship between wildfire soot and melt, in theory the heat-absorbing soot could fuel melt. Scientists have documented melt from the same wildfire-soot-induced darkening of glaciers in Canada.

Elsewhere in the far north—in Alaska, Canada, and Russia—scientists hypothesize that so-called zombie fires are likely on the rise as the climate warms. Those fires burn on tundra, and when winter comes, they continue to smolder underground. Like the undead, they never quite go out. Those areas of the Arctic also hold massive reserves of methane—a greenhouse gas even more potent than carbon dioxide—that could be unleashed into the atmosphere. Zombie fires "may have severe implications for fire management, human health, and climate," the authors of a 2021 study on the phenomenon argue.

Forests burn, releasing carbon into the atmosphere. Zombie fires unleash methane reserves. The relationship between wildfire and climate change is full of positive feedback loops, where one effect amplifies the next—a vicious cycle. Scientific consensus has landed on a couple of key figures when it comes to climate change: 1.5 degrees and 2 degrees Celsius warming above preindustrial levels. With 1.5 degrees comes more extreme heat and drought, flooding and displacement of people due to sea level rise, coral reef die-off, decreasing biodiversity, devastating shifts in crop and food production, and more. And 2 degrees would amplify those impacts even more. It's the line, scientists say, we don't want to cross.

Currently, the globe isn't on the best course. A UN report from fall 2022 says there's about a fifty-fifty chance the world will hit 1.5 degrees of warming, at least temporarily, in the next five years, and the most recent five years have been the warmest on record. According to the report, weather, climate, and water-related disasters have increased by a "factor of five" over the last fifty years. Those events—which include wildfire—cause about $202 million in economic loss every day.

Along with all this come larger and more destructive wildfires. The World Meteorological Organization predicts that extreme wildfires are likely to increase globally by 50 percent by the end of the century.

"We fire scientists have predicted an increase in wildfire activity and extreme wildfires out in the future, but not this soon," says Balch. "And so to see it happening this fast is really concerning. And you know I'll talk about it in a calm way, because I'm a scientist and I'm trying to be as objective as I can. But this is me hitting the panic button right now."

Whatever happens next is a major plot point in the long drama between humanity and fire. As people burned and burned, we shaped the world to fit our needs and desires. The world we have today was built on fire. But as that relationship got more complex, our mastery of fire escaped us. Through both wildfire and climate change, flames will shape the world that's possible in the future.

Stephen Pyne calls the modern era the "Pyrocene," a take on the Anthropocene, a proposed geological epoch defined by humanity's impact on the natural world. To Pyne, the Pyrocene refers to humans' ecological signature: the manipulation of fire. "We have created a Pyrocene," he writes. "Now we have to live in it."

WILDFIRE IN THE EARLY U.S.

To understand how to change the way we think about wildfire, we have to understand how we got here in the first place.

Before the West was colonized, wildfires would have burnt millions of acres a year. Wildfire acreage likely peaked around 1000 C.E., during a time of both high temperature and drought, and then again in the 1800s. One study suggests that back then, between 4 and 11 million acres of land burnt in California alone every year. For some perspective, as of 2022 the largest area of land that's burnt in a single year across the country since recordkeeping began is 10.125 million acres in 2015 (although two years since have nearly beat out that number). Compared with some of the natural conflagrations in the country's vast forests, the acreage burnt in even the biggest fire years of the last couple of decades wouldn't chart.

In addition to "naturally" burning fire, people have affected how and when fires burn for millennia. Before Europeans settled North America, there was scarcely a landscape or ecosystem not affected by human-influenced fire. Yurok and Karuk people, the Confederated Salish and Kootenai Tribes, the Lakota, Miwuk,

THE MOUNTAIN PINE BEETLE

The mountain pine beetle is about a quarter-inch long and looks almost like a tiny hippo with the legs of a shrimp. The beetles reproduce in lodgepole, ponderosa, limber, and whitebark pine. When the air warms enough in spring, they burrow through the bark of trees, lay eggs, and inject a fungus that both feeds their broods and blocks the trees from transporting water and nutrients. As they bore into trees, the beetles release pheromones that attract hundreds more insects to the same tree. The larvae feed on the fungus and on the tree's phloem, or the layer just beneath the bark. In only a few weeks, the beetles can destroy a tree's life support system. In a matter of months, the tree is dead.

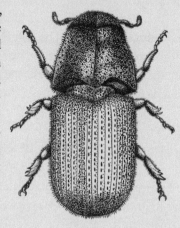

While beetles—like fire—have coevolved with our forests for millions of years, the Forest Service calls mountain pine beetles "the most aggressive, persistent, and destructive bark beetle in the western United States and Canada." Since 2000, mountain pine beetles have affected an area of forest in the western US nearly five times bigger than the entire state of Massachusetts. And this all links back to climate change.

Hot, dry summers make trees particularly susceptible to attack. Cold snaps can kill off infestations, but climate change is leading to milder winters. As temperatures rise, the beetles are expanding their range: both north—higher and higher into Canada—and upward in elevation. Decades ago, the limits of beetle attacks were around nine thousand feet. Now, infestations occur at eleven thousand feet. The mild temperatures are also accelerating the beetle's reproduction.

The beetles are entwined with wildfire as well. Some scientists and foresters argue that "red-stage" beetle-killed pine trees are particularly flammable. At that stage, pine needles glow almost the color of fire itself and are drained of their moisture. The logic goes: those dry needles are ripe for a burn. Studies show that this isn't so cut-and-dried. One 2011 study in the Yellowstone National Park area, for example, found that the probability of fire in the canopies of trees, often the fastest-moving, most devastating form of wildfire, went down for at least thirty-five years after a beetle outbreak. That's because the fuel in the crown of trees—pine needles, and then the trees themselves—fell to the forest floor as trees died from beetle infestation. But other studies shoot back: under certain conditions, in both experimental settings and real-world fires, fire severity *did* go up after beetle attacks, especially during the red stage.

Do bark beetle attacks make forests more flammable? The short answer is: it depends. Scientists do agree that the impact of the insects on fires is relatively negligible compared with the bigger picture of fire weather: drought, warm temperatures, and wind. However, while beetles may not have a huge influence on wildfires, wildfires might affect where beetles attack. After low- to moderate-severity wildfire, injured trees might attract beetle infestations, causing even more arboreal death.

Nez Perce, Ojibwe, Navajo, Apache, Blackfeet, ancestral Pueb-loans, and many more put fire on the landscape. Fire was at times used for battle or escaping threats. Other times, it cleared land or flushed game for hunting. Indigenous people used it to man-age land for specific types of plants and for ceremonies. Fire was sacred. It was a tool. It was a part of life.

Flames were a part of lived experience for settlers in all parts of the country too. Wildfire routinely sprung up in the West, on the plains, and in the East. In the 1840s, before moving to Walden Pond, the famed conservationist Henry David Thoreau started a three-hundred-acre fire in Massachusetts while cooking some fish he'd caught for dinner. He felt guilty at first, but later wrote in his journal that "it was a glorious spectacle, and I was the only one there to enjoy it." Thoreau recognized that fire was a natural part of the landscape—"I have set fire to the forest, but I have done no wrong therein," he penned. "These flames are but consuming their natural food." For years, residents of his town taunted him for the incident, calling him "woods burner."

A few decades later, Mark Twain wrote an account of a camping trip to Lake Tahoe in the early 1860s. After a long day out, Twain readied some bread, bacon, and coffee, set a fire, and went back to his boat for a frying pan. Suddenly, he heard his

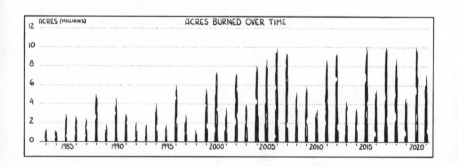

While the overall number of wildfires has remained about the same over the last four decades, the total area burnt has skyrocketed in that same timeframe.

friend yelling. He turned and saw the fire "galloping all over the premises." Like Thoreau, Twain was in awe of the force he'd set in motion. "It was wonderful to see with what fierce speed the tall sheet of flame traveled!" Twain wrote. Within thirty minutes, the fire had exploded. It burst up surrounding hillsides, onto ridges. Twain and his friend Johnny boarded their boat to escape the heat. "Every feature of the spectacle was repeated in the glowing mirror of the lake! We sat absorbed and motionless through four long hours."

At the same time, large fires were threatening lives and livelihoods, burning in massive spectacles across the country. In 1825, fires burnt an estimated 3 million acres in Maine and New Brunswick. In October 1871, fires raged to life across the Great Lakes. On the same day as the Great Chicago Fire, a burn in Peshtigo, Wisconsin, ravaged over a million acres and killed over a thousand people. Another burn, fueled by the same weather pattern, ignited 2.5 million acres in Michigan. The Hinckley Fire in Minnesota killed more than four hundred people in 1894. Over the next decade, fires in South Carolina, Oregon, Washington, and New York would burn through millions more acres.

By the early 1900s, fires were immensely destructive, but also natural, useful, even awe-inspiring. As settlers expanded westward, more changes were afoot that would make lasting shifts in our relationship with fire.

THE BIRTH OF PUBLIC LAND

Today, about 640 million acres of public lands are federally protected across the United States—more than a quarter of all land in the country. They are owned by the people, protecting water, wildlife, habitat, and entire ecosystems. From the canyonlands of Utah to the redwood forests of California to the marshes of the Everglades, public lands are a defining feature of America's natural and political ecosystems.

Those lands, though, have a contested and controversial history. In many ways, the origin of fire suppression is bound up with the origin of public lands in the U.S. To protect those lands—for the good of all, not just a property owner—would require coordinated effort, strategy, and resources.

In 1800, about 5.3 million people lived in the country. By 1910, the U.S. had a population of over 92 million, growing almost twentyfold in little more than a century. At the same time, people had pushed west. The spirit of so-called manifest destiny had dictated the killing and ousting of Indigenous peoples from their ancestral land. Many envisioned the country's bountiful natural resources to be limitless, but those fantasies confronted a dark reality, as settlers' manipulation of natural resources and wildlife began to get out of hand.

By the beginning of the twentieth century, predators like wolves, grizzly bears, and mountain lions were systematically exterminated and nearing extinction. Once numbering in the tens of millions, the last handful of bison was hanging on by a thread in Yellowstone National Park and a New York City zoo. Passenger pigeons—which once darkened skies with their numbers—were a rarity.

Extractive industry dominated the West. The U.S. government had given railroads millions of acres of land as right-of-ways to crisscross the country, expanding industry and trade. Timber and mining burgeoned in the West. Titans of those industries served in Congress or bought off politicians.

To some, the country's abundant natural resources—which once seemed endless—were under threat. Enter Gifford Pinchot, the lanky and mustachioed hero of public lands and national progenitor of conservation, who considered wildfire an enemy. Pinchot grew up a wealthy easterner. His grandfather had made his fortune clear-cutting forests. Before Pinchot left home for Yale, his father approached him and asked if he'd like to be a forester. At the time, the profession did not yet exist in the U.S., but with a love for the woods, and for the new and strange, he decided his future path.

After Yale, Pinchot went to France to learn forestry. There, trees were grown like a crop, and locals were prevented from lighting campfires. There was too much order and tidiness. He couldn't wait to get home. When he returned, he headed west. He saw the desert, the Sierra of California. He felt a sense of home. But his trip came to an end. Sullen, he returned to New York. Forestry, for Pinchot, was still just a dream. He hung a sign outside his office: Consulting Forester.

Pinchot was long considered one of the most eligible bachelors of the upper crust. According to one writer, "His eyes do not look as if they read books, but as if they gazed upon a cause." In combating the forces destroying the wilds of the West—the timber barons, railroad czars, mining magnates—he'd found one. "The exploiters," he wrote, "were pushing farther and farther into the wilderness." He saw other enemies too. "Of all the foes which attack the woodlands of America," he wrote, "no other is so terrible as fire."

By the dawn of the twentieth century, the nascent idea of public land had begun to bloom. In 1891, thirteen million acres had been set aside as "forest reserves," a precursor to what would become national forests.

In time, Pinchot became boxing, wrestling, and shooting buddies with Teddy Roosevelt. Pinchot wrote some of Roosevelt's speeches. Skinny-dipping in rivers, walking trails, the two went back and forth about their ideas of conservation. They both shared a dream of shifting the Republican Party away from serving the interests of a wealthy few to serving the American people as a whole.

In 1901, President William McKinley was assassinated. Roosevelt, then vice president, took the highest office in the nation. He had an opportunity to turn his dream into reality—and he wanted Pinchot to help him. "The forest reserves should be set apart forever for the use and benefit of our people as a whole and not sacrificed to the shortsighted greed of a few," Roosevelt said in his first address to Congress. Pinchot concurred. In his memoir, he wrote, "The

Conservationist and first head of the U.S. Forest Service Gifford Pinchot (left) and twenty-sixth president of the United States Theodore Roosevelt (right).

earth, I repeat, belongs of right to all its people, and not a minority, insignificant in numbers but tremendous in wealth and power."

Where writers like Thoreau and John Muir had extolled the virtues of preserving nature and wilderness, Pinchot saw the innate value of the natural world, but he also recognized that it held immense value for people. Nature was not something to be cordoned off and left untouched. It was to be protected for use by future generations.

The timber and mining industries out West—and their cronies in Congress—were furious at Roosevelt and Pinchot. A nascent battle was coming to a head: between those making a killing on the land and those who wanted to protect it. In 1904, Roosevelt won the presidency in a landslide. In one of his first acts as president, he created the U.S. Forest Service and named Gifford Pinchot as its first chief. Unlike the National Park Service or the U.S. Fish and Wildlife Service or the Bureau of Land Management—which fall under the Department of the Interior—the

Forest Service falls under the U.S. Department of Agriculture. Trees were managed as a crop, for human use.

A 1905 "Use Book" detailing what sorts of activities ought to be promoted and eliminated in the country's forest reserves described fire protection as one of the highest priorities. Loss from fire, the book said, "is beyond all estimate"—and putting a stop to that loss was up to the government. Laws had already been passed that could cost people $1,000, a year in prison, or both, for building a fire and leaving it before it was extinguished. Anyone who "willfully or maliciously" set fire to public land would get a two-year sentence, owe $5,000, or both. Still, the "Use Book" encouraged foresters to employ "utmost tact and vigilance" when dealing with settlers who had grown "accustomed to use fire in clearing land." It cautioned about further stimulating the already growing resentment against the federal government.

Roosevelt continued to set aside large tracts of public land. Much of that land was historically occupied by Indigenous peoples. From 1905 to 1909, the country's system of national forests grew by nearly one hundred million acres. During this period, government policies led to Indigenous tribes losing about eighty-six million acres of land across the country. The rise of the country's public land dovetailed with—and even necessitated—the systematic and often violent removal and erasure of Indigenous presence on the landscape. For example, by 1905, 98 percent of the Karuk people's ancestral land was under Forest Service control. That meant that the millennia-old Indigenous practice of burning would no longer be allowed.

But in the early 1900s, a legion of Forest Service rangers were out on the land, enforcing who could do what and where on federal land. The squadron quickly earned nicknames: "sissies," "Teddy's green rangers." Roosevelt's own party pushed against him. In response to Roosevelt's crusade for public land, his enemies tacked an amendment on a 1907 spending bill. It took away the president's ability to create new national forests without the consent of Congress.

The clock was ticking. Roosevelt had a week to sign the bill or risk a government shutdown. Pinchot saw a path forward. The journalist Timothy Egan describes Pinchot's thought process in *The Big Burn*: "Why not use the seven-day window to put as much land into the national forest system as possible?" Pinchot and Roosevelt set to work, poring over maps blanketing the floor of a room of the White House. By the end of the week, he'd added sixteen million acres of land to the national forest system.

For the first time, conservation entered politics and popular conversation. Over the course of his presidency, Roosevelt set aside 230 million acres of public land, swaths of timber and desert and mountains that were not to be owned by any company or individual—but by the American public.

In the early years of the Forest Service, not much land burned in forests. To Pinchot, that proved the agency and its rangers were capable of controlling fire. A prime duty of early rangers—along with trail building and surveying—was putting out fires. There was no professional discipline of firefighting. All knowledge was based on firsthand experience. No one knew it better than those rangers. But Pinchot was confident in their ability to stop the flames. "Today we understand forest fires are wholly within the control of men," he wrote.

This profound new idea—large-scale conservation of public land—had been put into practice but was still facing immense battles in the halls of Congress and for public sentiment.

"The conservation movement had grown from a series of disjoined efforts into the most vital single question before the American people," Pinchot wrote. Pinchot needed a story, an event that could catalyze the country into seeing the value of public ownership of public land. That event would come blazing down the steep mountainsides of the Northwest in the summer of 1910.

THE BIG BURN

By 1910, timber and mining interests in Congress had nearly suc-
ceeded in crippling the Forest Service. The service was in financial shambles. Rangers were being paid out of the personal pockets
of higher-ups. They were worn out and beat down. According to
Egan's masterful account of the era, a single ranger was responsible
for about three hundred thousand acres on average.

Like vultures circling dying prey, senators and newspapers
owned by industry sensed the agency was dying. One op-ed said,
"The Pinchot troop of foresters now infesting the West should be
called in, paid off, and abolished."

As the year unfolded, the Northwest entered historic drought,
setting the stage for a nasty fire season. Wildfires came early, starting in April. As the months wore on, dry lightning storms rolled
in almost every afternoon, setting off myriad small fires. Rangers
went out to wrestle the small blazes breaking out nearly every day.

By mid-August, more than seventeen hundred fires were
burning in Montana, Idaho, and Washington. Some of those
were accidental burns started by campers, westward travelers, and
homesteaders. Some were likely started by angry settlers, trying
to clear land for themselves or destroy the newly minted forest
reserves. At least one hundred of the fires started from coal-pow-
ered trains, spewing hot ash into bone-dry forests. One of Pinchot's rangers wrote, "All of nature seemed tense, unnatural, and
ominous." Things were ready to blow.

Then, in mid-August, walls of wind known as "palousers"
hammered the region. The sound, some said at the time, was like
that of a freight train. The winds reached over seventy miles per
hour; Egan calls them a "battering ram of forced air":

> The wind took the hot floor of the simmering forest and threw
> it into the air, where it lit the boughs of bigger ponderosas and
> white pines, which snapped off and also rode the force of upward

acceleration. Pine sap heated quickly and hissed as it reached a boiling point. Every headwall, every dead end of a canyon, every narrow valley served as a chimney, compressing the fire-laden air into funnels of flame.

In short order, those thousands of smaller fires converged and exploded.

People on the ground at the time described a red glow in the sky. A Forest Service ranger who experienced the blowup wrote, "Banners of incandescent flames licked the sky. Showers of large,

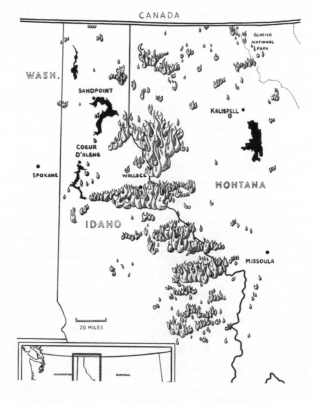

The Big Burn torched three million acres in a mere two days and played an enormous role in catalyzing the country's war on wildfire.

flaming brands were falling everywhere." Ed Pulaski, who later became legendary for his efforts to save his crew of forty-five men, wrote, "The whole world seemed to us men back in those mountains to be aflame. Many thought that it really was the end of the world." The winds were hurricane force. The sound of crashing timber was deafening. There were accounts of bears, deer, mountain lions sprinting in one direction "as if pursued by a foe that struck a deeper terror than man." Thirty years later, one writer penned, "Men who went through the fire were dumb for days in their struggle for words to measure the fire's horror." Another forester who fought the blaze wrote, "The spectacle of this fire was awe-inspiring almost beyond belief and was sufficient to strike terror to the strongest heart; it seemed a hopeless task to attempt anything that would divert the path of so ruthless a monster; many grown men in the crew were absolutely helpless and there were several who could only weep or moan, believing they were doomed."

For the first time, the government tried to organize to tame the flames. It was the first great battle of the U.S. Forest Service against fire. In the effort, the government freed convicts from local jails. They sent army battalions. Ten thousand people fought the flames. But, against a burn of over three million acres, they didn't stand a chance. Nature had overwhelmed the efforts of humanity. According to the Forest Service, the conflagration killed enough trees to fill up a freight train with wood that would stretch from Las Vegas to New York City. By the time the fire was reduced to a simmer, eighty-six people had died.

Contemporary accounts were apocalyptic and warlike. "Refugees" evacuated Wallace, Idaho, by train, heading for safer ground. The order was women and children first, but some men tried to shove their way onto the safe haven. Smoke blanketed much of the country. In some areas, one account said, "sun and daylight were shut out so completely as to cause artificial light to be used by day."

Coverage of the fire blanketed newspaper headlines throughout the country. The fire was called the Great Burn, the Big Burn, the Big Blow Up—all names getting at its size and ferocity. There had been deadlier fires in the past. There had been fires that burned nearly as big. But this one seized the country's attention. It was a clear battle, with high stakes: millions of dollars of timber and thousands of lives. If we think of the modern era of wildfire as a century-long drama, the Big Burn was the event from which everything else would unfold.

U.S. FIRE POLICY RISES FROM THE ASHES

By the time the Big Burn raged, Roosevelt had left office and Pinchot was engaged in a bitter feud with Roosevelt's successor, William Howard Taft. Taft, Pinchot thought, had strayed from the mission of public lands and fallen into the pockets of the robber barons he'd been fighting for years. Taft and Pinchot's battle escalated, and Pinchot was especially incensed with the newly appointed secretary of the interior, whom he viewed as an enemy of conservation. Their battle blew up over the use of coalfields in Alaska. By the time those thousands of individual fires converged into the massive blaze that seized the imagination of the country, Pinchot was out of a job. Taft had fired him from the Forest Service. But his influence remained.

The Big Burn became a crucial component of Pinchot's public relations campaign. He was hell-bent on making sure that his model of conservation endured and that the Forest Service would have a future—with fire suppression as a crucial component. For Pinchot, the nearly one hundred men who died fighting the massive fire had given their lives for this mission. He wrote op-eds, lobbied politicians, and gave speeches. "There were too many fires and too few rangers," Pinchot said after the burn. "The lesson from these forest fires is perfectly clear. When a city suffers from a great fire, it does not retrench in its fire department, but strengthens it. That is what the nation must do on the national forests."

Many people across the country remained vehemently against the Forest Service. One Idaho senator called the Big Burn "God's

Will." People viewed the immense destruction as a natural instrument of clearing room for more settlers. The Big Burn became fuel for another fire: a great culture war over the future of the country's public land. But traveling the country with Teddy Roosevelt, Pinchot leveraged the fire to make heroes of forest rangers, to make the case for more funding for the agency, and to make an enemy of wildfire.

The traveling pair found a receptive audience. The year after the Big Burn, Congress doubled the Forest Service's funding. In 1911, it passed a bill, called the Weeks Act, that allowed the federal government to purchase private land for protection in the East and collaborate with state and local organizations to fight fire.

In the battle for the hearts and minds of the American public, Pinchot had won. The Forest Service was here to stay.

SUPPRESSED

While the Forest Service's future had firmed up, its relationship with fire was not yet set in stone. In the wake of the Big Burn, a long-running debate picked up steam. Foresters, timber harvesters, and other people working and living in the forest were asking: Should we eliminate all fire, or are there cases in which the country should "let it burn"? Is fire always an enemy, or are there times in which it is a necessary, even vital, natural process? The outcome of this debate would shape the country's fire policy for decades to come and set in motion some of the factors leading to the increasingly catastrophic fires we see today.

The official policy of the Forest Service was clear: fire was an enemy to be vanquished. "That the fire menace is a real one needs no emphasis," Ferdinand Silcox wrote in 1911. At the time a district forester, Silcox went on to become chief of the Forest Service in the 1930s. In 1913, Pinchot's buddy and successor as head of the Forest Service said that "the necessity of preventing losses from forest fires requires no discussion. It is the fundamental obligation of the Forest Service." Two years later, the secretary of agriculture codified that sentiment, writing that the agency's primary duty—even over producing timber—was fire protection. By then, the Forest Service had improved trails and roads, built fire lookouts, and fine-tuned communication methods necessary to suppressing fires.

Indigenous people had been using fire across the country for millennia. For them, fire was a tool. It was at times sacred. It was a gift that could breathe life into the land. Some white settlers, often influenced by Indigenous practices, had taken to lighting their own fires. Proponents of the technique called it "light burning," and they did it to clear land, promote the growth of harvestable timber, and create habitat for game. Eight years before the Big Burn, an op-ed in a San Francisco newspaper denounced full-scale fire suppression because it allowed the accumulation of fuel for fires. In a 1910 issue of a magazine called *Sunset*, a timberman named George Hoxie, from Shasta, California, wrote that light burning should be mandatory. "Fires to our forests are as necessary as crematories and cemeteries to our cities and towns," he penned. To Hoxie, fire exclusion meant forests might as well be doused in gasoline; the overstocked, overgrown fuel without any natural control would be ready to go up in flames.

A debate over using fire as a tool began to take shape. In a 1919 speech, Joseph Kitts, a civil engineer and landowner, argued that the key to prevent devastating fire was: more fire. He said settlers had found forests open and clean when they arrived on the land, thanks to Indigenous burning. Those fires had cleared away brush, dead trees, and branches. Indigenous people, he said, were "the most practical of foresters." That same year, the land commissioner for the railroad corporation Southern Pacific Company said that the Forest Service's suppression policy meant that "there will come a day when a forest fire will sweep the entire coast from Mexico to Canada and leave little or no vestige of our existing forests." He contended that the forests of the West had evolved with fire and proposed that the Forest Service adopt light burning.

The Forest Service wasn't having any of it. By 1920, the debate had gone mainstream and was coming to a head in California. The Forest Service began threatening magazines promoting light burning with lawsuits. Assistant Forester William Greeley, soon to be promoted to head of the Forest Service, denounced the practice as "Piute Forestry." Harnessing the racism of the day, he depicted light, controlled burning as unscientific and primitive, contrasted with the professional know-how of the agency's highly trained foresters. The Forest Service's bottom line at the time was timber production. "If the only solution lies in the uninterrupted destruction of young growth by light burning, we had better harvest our mature stumpage without more ado and become a wood importing nation," he wrote facetiously. He called for a "real program of fire protection," along with "harder and more united efforts by all agencies, public and private."

The Forest Service even began conducting its own studies, with the explicit intent of demonstrating the failures of light burning. Researchers rigged the study; they placed fuel like pine limbs along the trunks of trees, to ensure that they burnt in the fires. But the Forest Service's voice and influence was strong. By the mid-1920s, light burning was heresy in California. But the call for careful, controlled fire wasn't yet dead.

In the Southeast, burning continued, and clashed with the dominant mindset toward fire. But as northern landowners purchased swaths of land for hunting retreats in the South, they were appalled at the fiery landscape they saw. They put a stop to burning. They also noticed that their hunting opportunities were declining—in particular, they started talks with the government about why bobwhite quail populations were shrinking. The government hired a scientist named Herbert L. Stoddard to investigate. Stoddard concluded that fire exclusion in longleaf pine forests of the South had led to the decline of habitat. In the face of adamant and constant criticism, he argued that fire was key to the revival of the species. The book that came out of his investigations became a classic text in wildlife conservation.

These findings didn't budge the Forest Service's attitude toward fire. As Forest Service chief, Greeley tried to rein in Stoddard's writing and contended that annual burning of the woods depleted forests' stock of game animals, rather than restoring them.

By the mid-1930s, more wildfires had hit the West—like the 1933 Tillamook Fire, which burned three hundred thousand acres in Oregon. Even though foresters were still debating the value of controlled burns in conferences and papers, the Forest Service had made up its mind. Fire, the agency contended, was bad. And that's all there was to it. In 1935, Chief Forester Silcox put forth what would become a landmark standard for the Forest Service: the ten A.M. policy. That rule stated that every wildfire ought to be put out by ten A.M. the next morning. Total suppression became the name of the game, and the Forest Service became an organized front of anti-fire propaganda. The rogues and renegades—self-professed "heretics"—promoting light burning were a loose collection of passionate burners. Though they agreed that fire was crucial to forests, they couldn't always agree on how and when forests should experience fire.

The Forest Service believed that the public could comprehend only a single, unified message. Just like Pinchot's crusade for American public lands, the battle was framed as "good versus evil." For the Forest Service, the battle was all hands on deck. It withheld federal funds to state forestry in states that allowed light burning. A massive public relations campaign began.

The agency created children's songs, made pamphlets and ads, and sent legions of young foresters to the South to combat "woodsburning." It viewed the drive to start controlled fires as a mental illness of sorts. It even hired a psychologist to study why southerners kept burning, despite the federal government's best efforts at taming the flames.

It was a precarious moment. Had proponents of "light burning" won out, our relationship with fire could have mapped an entirely different path for the century to come. The irony here is that today fire experts call for controlled—or "prescribed burning"—on millions of acres of land across the country. The debate over light burning is a time capsule that shows what could have been. Had light burning caught hold in the West, would our forests today look different? Would our wildfire seasons simmer instead of roar? Was this a chance to establish a healthy relationship with fire that we passed up? Can we regrasp a tolerance of flame and smoke?

Instead, though, the mission of fire suppression reigned supreme. Fire was an enemy to be extinguished. And that no-tolerance strategy for wildfire had some unfortunate unintended outcomes.

As the twentieth century wore on, more people were living in urban areas than rural ones for the first time in American history. That meant most of the American public was dissociated from fire. They didn't burn slash or grass on their own land. They rarely breathed the smoke or saw the flames of conflagrations in forests. Fire was mostly contained to the chimney. Nearly all the country had been settled. Along with that, just about all major predators in the wild had been killed off. Mountain lions, wolves, and grizzly bears were all extinct in most parts of the Lower 48 states—or nearing it. Fire, for much of the country, was another beast to be eliminated. According to historian Stephen Pyne, "It was a crazy ambition."

By the early 1940s, the country's anti-fire campaign had taken on the literal strategies of war and wrapped itself in patriotism. During World War II, Japan sent thousands of incendiary balloons floating to the U.S. via the jet stream. Most of the bombs fell into the ocean before they could reach American soil. They didn't ignite any major forest fires, but six people were killed in a church group when they discovered one of the balloons in an Oregon national forest. The public grew worried that foreign bombs could explode in the West, igniting huge fires. At the same time, legions of men who might otherwise be working in the forests, fighting fires, had been sent overseas. The U.S. government had to figure out other ways of fighting wildfire.

A BEAR IS BORN

On October 10, 1944, the artist Albert Staehle delivered what would become the fluffy, ursine symbol of the country's obsession with fire suppression: the first rendering of Smokey Bear.

Eventually, Smokey also got real-life representation. In 1950, some crews on a wildfire in New Mexico's Capitan Mountains came across a black bear cub in a tree, suffering from burn wounds. The crews removed the cub, and it quickly gained fame. It traveled in a private airplane, with an image of the bear with its paw in a sling and signature Forest Service ranger cap on its head on the fuselage, to the National Zoo in Washington, D.C. When the plane touched down, hundreds of spectators were awaiting its arrival. The bear was named "Smokey," and its burn injuries left it with what the *New York Times* called a "stiffleged stomping gait." He became an icon who received thousands of letters a week and was granted his own zip code. Smokey lived to be twenty-six years old. The *New York Times*, *Washington Post*, and *Wall Street Journal* all published obituaries when he passed away. His remains were buried at Smokey Bear Historical Park, back in Capitan, New Mexico.

Nearly eighty years later, the Smokey Bear campaign is still running strong. Smokey has his own Twitter, Instagram, and Facebook pages. Smokey mascots attend farmers markets and fairs across the country. His face still adorns posters and ads. Smokey has become the longest-running public service ad campaign in American history.

The government took up a media campaign, tied directly to the war effort. "Careless matches aid the Axis—prevent forest fires," one early poster read. Other messaging was similar: "Don't blind our pilots with smoke—prevent forest fires." "When you're fighting fires you're not logging! Help on the home front to win the war. Keep production up." Wildfire fighting itself took on the organization, equipment, and precision of a military operation.

The government turned to another measure in the battle for America's hearts and minds too: advertising. After some experimenting with what message would land with the public, Smokey Bear sprang into existence. His slogan, "Remember...only you can prevent forest fires," echoes to this day. That catchphrase makes wildfire an individual problem, not a collective one. People camping and recreating in the forest, the logic goes, are responsible for our big fires. And they, too, can stop them. The country's anti-fire campaign endures. Today, 98 percent of fires are put out during "initial attack," when they're still a couple of acres or smaller.

SCIENCE COMES AROUND

By the late 1950s, wildfire was getting more attention in urban areas. The National Fire Protection Association called hilltop homes in Los Angeles, built amid flammable shrubs and trees, a "design for disaster." A few years later, a Forest Service researcher wrote that people occupying the flammable areas of California "increased the potential for conflagrations between two and ten times over the level of the 1930s." The first inklings of the fiery future were beginning to emerge.

Over the next few years another, potentially competing idea began to come forth: maybe fire wasn't so bad after all. In the early 1960s, the secretary of the interior put together a special advisory board on managing wildlife, headed by Starker Leopold, at the time a professor at the University of California, Berkeley. Leopold spent his youth in New Mexico while his father—the famed conservationist Aldo Leopold—worked for the Forest Service. With dark short hair, thick eyebrows, and prominent smile lines projecting from the corners of his eyes, Leopold long had a fascination with wildlife.

The report produced by Leopold's group, released in 1963 and later dubbed "The Leopold Report," would lay the foundations of how the National Park Service managed its ecosystems for decades to come. The Forest Service, under the Department of Agriculture—not the Department of the Interior—would take years longer to get the message. The report famously declared that the country's parks should preserve "a vignette of primitive America." The paper suggests that parks underwent "periods of indiscriminate logging, burning, livestock grazing, hunting, and predator control." Then, the paper argues, there was an abrupt shift. Suddenly there began "a regime of equally unnatural protection from lightning fires" and other natural forces. National Parks, the report argued, are areas where nature should, by and large, be left to nature.

To restore nature, the report contended, "the controlled use of fire is the most 'natural' and much the cheapest and easiest to apply." But to reintroduce flames, forests would need careful preparation and treatment. Trees and shrubs might need to be cut, piled, and burned—or else there could be risk of an enormous, out-of-control wildfire from the decades of unnatural growth that could serve as a fuel. The report didn't devote all that many words to fire, but the impact was sudden. It was the beginning of a fire revolution.

When the report came out, another scientist—also at UC Berkeley—was already laying the groundwork to prove that controlled fire could work. In his long career putting fire on the ground, Harold Biswell earned the nicknames "Harry the Torch," "Burn-em-up Biswell," and "Doctor Burnwell." He put the ideas Leopold proposed to the test on the ground in the Sierra Nevada.

In the mid-sixties, Biswell started working on a plot of land right next to Sequoia and Kings Canyon National Parks. Sequoias—enormous, majestic conifers that can stand taller than the Statue of Liberty—were in decline. In 1969, his team thinned the "unnaturally" overgrown forest and burned about one hundred acres to see what it might mean for the iconic trees. They found that, after the burn, there were tens of thousands of seedling sequoias in the burned areas. There were none in the control area that hadn't been burnt at all. That's because sequoias *need* fire. The heat and flames trigger a reaction in the trees. High up in their canopies, small green cones sit waiting to burst and send their innards scattering to the soil below. A single cone averages about two hundred seeds, and it can store them for twenty years. When fire comes, the heat dries out the cones and cracks them. The seeds erupt into the air, allowing more sequoias to grow. Without periodic fire, the sequoias cannot reproduce.

As he was actively burning, Biswell also conducted "field days" to show students, foresters, and the public that fire wasn't all bad. Fire historian Stephen Pyne says Biswell's troop of students at UC Berkeley did for fire "what Yosemite's Camp 4 did for rock climbing." These renegades, on the fringes of mainstream science, brought innovation that pushed the boundaries of accepted knowledge about ecology. Biswell, his colleagues, and his students showed that fire wasn't just good for a healthy landscape. It was a necessity.

Sequoia and Kings Canyon National Parks started policies that allowed prescribed burning and natural fires to simmer unimpeded in the late 1960s. By 1970, similar ideas had spread to Yosemite too. Biswell was instrumental. In the years to come, prescribed fire was embraced in many California state parks, as well. By 1978, prescribed fire was also allowed in national forests.

Scientists and foresters were recognizing that the country's decades-old no-tolerance policy for wildfire had some nasty side effects: forests were overgrown, trees and plants that depended on fire were deprived of the nourishment they required, and conditions across the country had departed from the "natural" state they'd been in when the continent was settled. The gospel of fire was spreading, fast.

In his memoir published in the 1960s, Stoddard, the quail researcher, wrote that the anti-fire endeavor was "the most intensive—and ludicrous—educational campaign that ever insulted the intelligence of American audiences. It was carried on by well-meaning but utterly misinformed persons."

As this revolution in wildlife, ecology, and fire science was happening, other changes began to take hold of the American public. The country was undergoing a revolution in environmental values. In 1962, the scientist and writer Rachel Carson published *Silent Spring*, a landmark work in the environmental movement. In the first chapter, she describes a fictional town in the heart of America. "There was a strange stillness," she writes. "The birds, for example—where had they gone? Many people spoke of them, puzzled and disturbed. The feeding stations in the backyards were deserted. The few birds seen anywhere were moribund; they trembled and could not fly. It was a spring without voices." Though the specifics were fictional, Carson's opening parable was meant to detail what was happening across the country. Springtime was growing quiet, thanks to the damage humans had wrought on nature. In particular, the pesticide DDT—sprayed en masse across the country in the name of insect and pest control—led to paperthin eggshells in bird populations, decimating avian populations. Human intervention had thrown nature out of whack, and the public was beginning to notice.

The book seized the country. In a subcommittee hearing on pesticides a year after the book's publication, Ernest Gruening, a senator from Alaska, told Carson, "Every once in a while in the history of mankind, a book has appeared which has substantially altered the course of history." *Silent Spring* was one of those books.

In 1964, Congress passed the Wilderness Act, designating more than nine million acres of so-called wilderness in national forests across the country. Today, that number's grown to over one hundred million acres—an area about the size of California. The idea of wilderness was to preserve areas free of the impact of humans and development. Wilderness, the act proclaimed, was "an area where the earth and its community of life are untrammeled by man, where man himself is a visitor who does not remain." At the act's inception, these ideas erased the Indigenous history of the country's wilderness areas. For thousands of years, people had been living in and shaping even America's "wildest" locales. But the act was a radical change for huge swaths of public land and reflected a swelling sentiment that land ought to be *preserved* for its own sake—not *conserved* for the use of people.

More works of environmental journalism and fiction followed, reflecting the country's growing fascination with wildness. In 1968 came Edward Abbey's *Desert Solitaire*—a seminal tirade in part against the "industrialization" of national parks and in support of the harsh beauty of desert environments. Seven years later, he published *The Monkey Wrench Gang*, a rollicking book about what some call ecoterrorism: putting nails in trees to thwart logging, plotting to blow up Glen Canyon Dam, cutting down roadside billboards. It inspired the formation of the radical environmental group Earth First, and today "monkey wrenching" is synonymous with ecosabotage.

The first Earth Day was held in 1970. In the years to come—during the Nixon administration—the Clean Air Act, Clean Water Act, Endangered Species Act, and other legislation protecting habitat, wildlife, and ecosystems were all passed with overwhelming bipartisan support. An environmental reckoning was taking shape. The idea that humanity could inextricably and irrevocably destroy, manipulate, and use up nature was going mainstream. Sometimes, the public was demanding, nature ought to be left alone to its own devices.

"If we date the origins of modern, American fire history from 1910, half of that history has been spent trying to take fire out of the landscape and the last half has been trying to put it back in," Pyne says. "It turns out, however, that it's a lot easier to take it out than to put it in."

It was this shift in values—not just in science, as put forth by Starker Leopold and Biswell—that helped prescribed fire and tolerance for natural burns take hold. Then, in the late 1980s, came an event that would put the new science of returning fire to the land to the test.

THE YELLOWSTONE FIRES

Fires had burnt in and around Yellowstone National Park since the area's glaciers retreated about twelve thousand years ago. But in 1988, people across the country hadn't seen a huge burn in the area in living memory. The spring of that year in the region was wet. That moisture allowed grasses and undergrowth to grow tall. Then, starting in June, that moisture dried up. In July, August, and September, the region entered its worst drought on record.

During that summer, the National Park Service allowed eighteen lightning-started fires to burn. They were natural fires that could theoretically be good for the ecosystem. Eleven of those fires burned out with no hoopla. As the drought hit, park officials realized just how flammable the landscape was becoming. By July 15, they didn't allow any new fires to burn that were outside those already existing paths of flames. A week later, the park began to aggressively fight the spread of the flames. Within another seven days, the fires within the park were burning about 100,000 acres.

Nearly 250 wildfires started in the region that summer. Forty-five of them were within Yellowstone's boundaries. About 1.3 million acres burnt across the greater Yellowstone area. Thirty-six percent of the park had gone up in flames. About half of the total acreage burnt was from human-caused fires.

As the fires exploded, the country mobilized the largest fire suppression effort it had ever seen. More than twenty-five thousand firefighters were deployed to the area at a cost of nearly $120 million.

Flames flew as high as two hundred feet, shooting embers into new areas. Weather conditions meant that the fires were nearly impossible to combat. One incident commander said, "We tried everything we can think of and that fire kicked our ass from one end of the park to the other."

On August 20, extreme winds hit the region. The fires spread another 150,000 acres. The day became known as "Black Saturday." Reporters flocked to the area. The severity of the fire and what it meant for the region depended on whom you were listening to and reading. Especially in East Coast outlets, the park was sometimes reported to be a "blackened moonscape." Misinformation spread as fast as the flames. Many people were under the impression that the park's pro-fire policies meant that park officials weren't doing anything to combat the blaze.

The situation became political dynamite. Ronald Reagan reportedly called the park's let-it-burn policy "cockamamie." Other politicians were equally critical. In politics, there was no room for scientific middle ground or nuance. The issue was crystal clear: fire was bad. To some—especially those far away from it—fire was still an evil to be eliminated. But the idea that ecosystems need fire was seeping deeper into the fabric of the agencies responsible for managing it. The challenge was finding the right balance of "good fire" and communicating how that might look to the public.

After the fire, studies began on what the burn meant for the ecosystem. Wildlife, it turned out, was by and large okay. Though some elk and other critters died from smoke inhalation, most animals simply moved to safe ground. At a conference not long after the fire, a University of Wisconsin–Stevens Point professor gave a talk on the ecology of fire. He said that if he happened into the park's superintendent position, "I would maintain fire every chance I had." He pulled up a slide depicting Smokey Bear, and gestured to the big, brown, jeans-clad ursine. "I would do my best to eradicate this species from the park." In an October analysis of the fire by the park, the authors wrote, "The most unfortunate public and media misconception about the Yellowstone firefighting effort may have been that human beings can always control fire if they really want to; the raw, unbridled power of these fires cannot be overemphasized."

Today, travelers to Yellowstone might not recognize signs of the fire at all—aside from a few interpretive signs. Native species returned to the landscape. Aspen flourished. About 80 percent of the park's forests are dominated by lodgepole pines that have "serotinous" cones. Like sequoias, those only release their seeds when superheated over 113 degrees. Fire was just the trick they needed. The trees are adapted to even big blazes like the 1988 fires—historically, large, "stand-replacing" burns would roll through every 150 to 400 years. Even though some of the fires were human-caused, an ecological assessment of the burns concluded that they were, by and large, within the range of natural, historical variability. The fires also generated a complex mosaic of habitat—far from the "blackened moonscape" of early reports. The sheer scale of the fires provided an unequaled natural laboratory for further study.

It's easy to say "Yellowstone has recovered." But Andrew Larson, director of University of Montana's Wilderness Institute, is skeptical of using "recover" to describe postfire landscape. That implies something bad happened to what was there before. He's studied disturbances like wildfire in forests for decades. Fire, he says, "it's not good or bad—it just is." Scientifically, he says the Yellowstone fires were a turning point. "It helped us realize that these major, widespread fire events were part of how ecosystems work."

Larson happened to be in Yellowstone on vacation with his family during the blazes. He was only a child, but he remembers seeing a small fire across a creek, before things blew up. "I remember having this intense emotional response," he says. "Nine-year-old me wanted so urgently to cross that creek and put that little fire out. I was just desperate to do it."

Larson's grandpa talked him out of that idea, telling him that sometimes it was okay for a fire to simmer like that. But Larson says that memory shows him that he had already been programmed to instinctively consider fire evil—like so much of the American public. Fire is such a visceral force that it's much easier to grasp in just one way than as something that has as much nuance as nature. That moment in his childhood is symbolic to Larson.

He worries that, after decades of science around the importance of fire for ecosystems and the futility of eliminating it from the landscape, society hasn't made any progress past that emotion he felt as a child. "Sometimes it makes me feel a little bit, I don't know... disappointed. Am I making any difference here with the science?"

The same tension that plagued the Forest Service from its foundation still exists today. How we resolve it will in large part inform the path we take in confronting the ever-more devastating wildfires of our future.

FIRE REGIMES

One way to look at fire history is to look at tree rings. When fire burns a tree, it triggers a response: sap pours into the "wound." Drilling cores in trees, scientists can analyze those rings and get a historical look at how often fires burnt through certain ecosystems. Working together, researchers have even formed an enormous database, consisting of over thirty-seven thousand fire-scarred trees from Alaska down to Mexico.

Using the data of tree rings, scientists can piece together a picture of the "fire regimes" across the country. Fire regime refers to how often and intensely fires burn in an ecosystem. Fire regimes interact with overall climate conditions and differ during periods of moisture and drought.

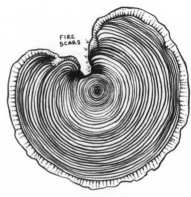

Tree rings depict the age of trees— and also when they experienced wildfires, a natural and crucial part of many species' life history.

FIRE WEATHER

Back in the late 1990s, Michael Fromm, a researcher at the Naval Research Lab in Washington, D.C., noticed something curious on some satellite data he'd been looking at. Layers of ash and soot in the atmosphere created the appearance of a volcanic eruption. But there had been no eruption to produce it. "We called them mystery clouds," Fromm says.

Fromm and his colleagues coordinated with researchers around the world who had made similar observations. Finally, they pinned the source of the ash on forest fires in Canada. Smoky skies were common during wildfires, but this smoke was just *so high up*, almost into the stratosphere.

As the years went on, Fromm and his team kept studying this phenomenon. Eventually, these extreme smoke clouds earned a name: pyrocumulonimbus clouds. Essentially, they're weather systems generated by extreme wildfires. The atmospheric conditions need to be able to permit the formation of a regular thunderstorm. Then, the intense heat of the fire creates an additional updraft. It generates its own cloud, and eventually what Fromm calls "a full-fledged thunderstorm."

While pyrocumulonimbus clouds are an ominous sight, they can also create a particularly destructive feedback loop. As the already raging fire generates its own weather, the lightning and intense winds associated with the fire thunderstorm can cause new starts in the area, accelerating how fast the fire spreads. Fromm says that some of the most destructive fires in history—like Australia's Black Saturday in 2009, which killed 173 people—have generated these clouds.

In many areas, studies of those tree rings show a century-long gap in fire. U.S. fire suppression policy became burnt into the skin of nature itself. Other causes led to the lack of fire too. Cattle and other livestock, for example, ate up grasses that would have served as fuel for the sorts of low-intensity fires ecosystems need. As human populations grew, they built roads and other areas devoid of fuel that served as fire breaks. But the upshot is the same: human intervention led to less fire over the last one hundred years or so.

This picture is especially true in low-elevation ponderosa pine forests. Fire regimes differ wildly depending on the area, but in most of those ecosystems, low-intensity wildfires would've likely burnt through every forty years, or in many cases much more frequently than that.

As those fires have been held off, small trees that would have ordinarily been burnt up have grown larger. Those low-elevation forests became thick and overgrown. Ponderosa pine is the main species on over twenty-seven million acres in the West. That's an area about the size of Tennessee. Ponderosas have a thick, corky bark and get more fire-resistant as they age. In many forests across the west, ponderosa pines have increased in density from 49 to 124 trees per hectare to 1,234 to 2,470 trees per hectare. Especially in those thick, dense forests, when a burn does ignite, it does so with a ferocity and scale that would rarely have occurred had fires in those ecosystems burnt naturally. Fires may sear the soil, making it difficult or impossible for new growth to come back.

Some call this part of fire ecology the "ponderosa pine story." Where fires were held off for the better part of a century, more fuel grew in and made forests ready to burn more intensely than they ever would have before. Where there would've been relatively benign fires crawling along the forest surface historically, modern fires might climb into the canopies, or crowns, of trees and burn entire stands.

It's easy to pin our predicament only on human intervention—specifically, fire suppression and a rapidly changing climate. However, nature rarely operates according to one storyline. For example, things look different up high. Higher-elevation forests see much longer intervals without fire. When fire does occur, it's generally larger, higher-intensity, stand-replacing burns. There, not as much so-called unnatural growth has sprouted up due to fire suppression. Century-long gaps in fire—or even much longer— are perfectly in accordance with the historical record.

In between low and high-up forests lies what some call the "messy middle." Frequent, low-severity fires hit sometimes. So did big, stand-replacing burns. In some areas, fires burnt large swaths of forests. Other areas burnt only a little. There's no single story for this area.

There isn't one large, unbroken canopy. There's a patchwork: some areas are incinerated. Other areas flourish. That pattern, often referred to as a "mosaic," also provides valuable habitat for all kinds of wildlife.

Thinking of fire as "unnatural"—or burning uncharacteristically aggressively due to human fire suppression and the buildup of fuels—makes it easy to pin that fire as a "bad fire," to justify a full-scale fire suppression effort, even with the knowledge that ecosystems depend on fire. But sometimes, fires need to happen— even if they're big. Sometimes they're just what a forest needs.

HOUSES ON FIRE

Part of the story of the twentieth century is the story of urbanization. As cities filled up and became denser, affordable vehicles helped people move to suburbs. Cheaper land outside the city meant more development on the fringes of urban places. The overall population was booming too. In 1950, the U.S. population sat at around 150 million people. By 2010, that number had more than doubled. Today, more people than ever live in the West—a boom in population partly fueled by the area's abundant natural areas and amenities. But along with that explosive growth came an unintentional side effect: more people than ever are living in areas prone to wildfire.

A few decades ago, wildfires from time to time hit communities. But that sort of destruction wasn't annual news, as it's become now. In 1970, for example, hundreds of wildfires ignited across Southern California. By the season's end, the flames had torched more than half a million acres. More than seven hundred homes burnt down and sixteen people died. At the time, that season was an outlier. But the destruction fueled an overhaul in how fires are managed nationwide.

Back then, the country's fire response system was tremendously different than it is today. Local areas had their own methods of management and structures of command. A study of the agency response to those destructive California fires deemed that there was poor planning, organization, and communication in response to the fires. The federal government responded by funding a program called FIRESCOPE, meant to remedy the situation and improve how future fires were tackled.

From FIRESCOPE came something called the Incident Command System, or ICS. It also birthed a system of coordination between agencies' fire response. By the 1980s, the program that started in California went national. In 1987, the Federal Emergency Response Agency adopted the Incident Command System. That meant it could be used to respond to all kinds of disasters—not just wildfires. In 1989, an Exxon oil tanker dumped 10.8 million gallons of oil off the coast near Valdez, Alaska. The response to that disaster became one of the first times the federal government adopted ICS outside wildfire. The terrorist attacks of September 11, 2001, further cemented that structure as a national model of managing enormous, destructive events. The ICS has been further honed, and today it provides the backbone of coordinating and deploying wildfire resources nationwide. As the ICS was forming, more homes were burning. In 1980, the Panorama Fire in Southern California burned three hundred homes. Five years later, about one hundred houses burned in Palm Coast, Florida. These events were still the exception, not the rule. However, concern was growing over wildfires consuming more and more homes. In the mid-1980s, the

WILDLAND URBAN INTERFACE

federal government started using the term *wildland urban interface*, which refers to areas where homes abut trees, shrubs, and other stuff that can burn. Of particular concern is the so-called intermix, where homes straight-up mingle with fuel for flames.

Despite those few, destructive fires, and the Yellowstone conflagration of 1988, not many people thought urban wildfires were much to worry about. Fires were consuming historically low acreage overall. Many people thought that the fire problem had been by and large solved. But they didn't see what was coming. In the early 1990s, a series of destructive urban fires hit California: the 1990 Painted Cave Fire in Santa Barbara burnt 430 buildings. In 1991, the "Oakland Hills Firestorm" torched 4,000 homes and apartments in the Bay Area. In 1993, a wildfire in Laguna Beach left another 400 homes in ashes.

The problem wasn't just that fire was burning more homes; it was that people were building more homes in wildfire-prone areas. Part of the debate over the WUI is actually a much older phenomenon, known in different terms: urban sprawl. As the twentieth century progressed, lack of zoning and other restrictions made living outside city centers cheaper. "Wilder" locales also had an aesthetic appeal; people loved living in and among trees and trails.

By the twenty-first century, WUI growth in the United States had gone into overdrive. Today, the WUI is growing faster than any other land type use in the country. More than a third of homes in the U.S. fall within the WUI, and that's almost entirely due to new development—not new growth in trees, forests, and wild areas. As of 2020, about sixteen million homes in the West are in the WUI. State by state, California towers above the rest of the country at about five million homes in fire-prone areas. In Colorado, there are about a million. Texas has a little over three million homes vulnerable to flames.

The escalating damage of wildfires reflects these trends. In 2011, for example, a devastating fire season in Texas left about four million acres burnt and nearly three thousand homes in ashes. California's Camp Fire in 2018 destroyed more than eighteen thousand structures. The 2020 Almeda Fire in Oregon swept through more than twenty-six hundred homes. Colorado saw its most destructive fire on record when the Marshall Fire, near Denver, destroyed more than one thousand homes in the last days of 2021. The 2022 Calf Canyon and Hermits Peak Fires in New Mexico left nearly one thousand homes in ashes. From 2005 to June 2022, wildfires have incinerated more than ninety-seven thousand structures.

Still, growth in the WUI continues almost unabated. More than half of homes built in the 2010s and 2020s face fire risk, compared with less than 20 percent in any decade before the 1970s. That number is even more striking when looking at state-level data. Ninety-seven percent of new homes in Arizona built since 2020 are at risk of wildfire. California sits at 91 percent. In Colorado, that number's 90 percent. Wyoming and Montana both hover around 93 percent. Utah is at 85 percent.

Kimi Barrett, a researcher who focuses on the WUI for nonprofit think tank Headwaters Economics, says it might take an enormous, destructive, transformational event to jar people into taking the WUI seriously. "It's pretty terrifying to think about," she says.

SMOKE AND HEALTH

Wildfire smoke is one of the most visible effects of wildfires. It can mean indoor recesses, canceled sporting events, and scrapped vacations. But smoke is more than just annoying: as the number of extremely smoky days climbs, scientists are learning more about what that could mean for the health of millions of people who are exposed.

Wildfire smoke contains what's known as "fine particulate matter," or tiny particles that can stick to lungs and enter the bloodstream, potentially taking years off lives. A 2022 Stanford University study found that wildfire smoke exposure has skyrocketed over the last decade. The study looked at where wildfire smoke shows up and how severe it is. More than fifteen million people breathe in unhealthy levels of smoke every year, they concluded. These findings suggest that a public health crisis is looming across the West.

Already, studies show that inhaling wildfire smoke is bad, especially for children and folks with breathing problems like asthma or emphysema. It also impairs our cognitive ability. Even short-term exposure can reduce attention span. Wildfire smoke is especially dangerous for developing fetuses. It can lower birth weight, increase the likelihood of premature birth and cleft palate, and contribute to gestational diabetes.

It's best to prepare for smoke now, so you're not caught off guard when it comes. On page 180, you'll find tips for minimizing your exposure and mitigating the health risks.

WHY THE WEST?

Nearly all natural landscapes in America evolved with fire. The Great Lakes region consistently sees wildfire, and scientists predict more is to come as the climate changes. Some of the deadliest fires in American history, like the 1871 Peshtigo Fire, occurred in Wisconsin. Fires burn in Kansas and Missouri, in Alabama, Tennessee, New Jersey, Florida, Alabama. So why does the West bear the brunt of the most catastrophic burns?

Part of the answer has to do with how the area was settled. In the East, many landscapes are more broken up by roads and development. In the West, many saw what unfettered development could mean and set aside a tremendous amount of public land. In the West, the federal government owns more than 50 percent of the land. In the East, that number is closer to 5 percent. The western half of the country has huge swaths of unbroken forest that can carry flame through massive areas.

Another part of the answer has to do with climate and precipitation. Back in 1878, the famed one-armed explorer and Civil War veteran John Wesley Powell had wrapped up two trips rafting the Grand Canyon. He'd hiked and floated through enormous sections of the western U.S., garnering public fame and scientific acclaim while doing so. He put forth an influential paper to Congress. In it, he described a dividing line that split the country in two.

That line—the 100th meridian—separated the relatively wet East from the dry, arid West. It runs through the Dakotas, Nebraska, Kansas, Oklahoma, and Texas. Drive through any of those states, and you might notice the phenomenon yourself. Here's how Powell himself described it:

> Passing from east to west across this belt a wonderful transformation is observed. On the east a luxuriant growth of grass is seen, and the gaudy growth of flowers of the order *Compositae* make the prairie landscape beautiful. Passing westward, species

100TH
MERIDIAN

after species of luxuriant plants and brilliant flowering plants disappear; the ground gradually becomes naked with "bunch" grasses here and there; now and then a thorny cactus is seen, and the yucca thrusts out its sharp bayonets.

Even in the late 1800s, Powell understood the fire dependence of the ecosystem. He wrote that fire holds back trees from growing on the prairie by burning away seedlings, and that fire is part of the ecosystem. However, as was the attitude of the day, he also advocated for its control—and the removal of Native Americans, who used fire on the landscape. That racism, built into much thought of the day, proved prescient in America's war on flames.

But to return to the question of why the West has so much fire compared with the rest of the country, the simple answer is: the West is drier, and less moisture equals more fire. In the West fires can burn hot and quick; they can also burn for months on end; and they can ravage millions of contiguous acres. But there's a twist: scientists are finding that climate change is pushing the line that divides the country farther and farther east. Some science suggests that that dividing line is now closer to the 98th meridian, nearly 150 miles east of the 100th. As the climate changes, the drought-ridden and dry landscape will continue to expand.

Even separating into "the West" doesn't do justice to the complexity of the problem. Fires burn differently in the high desert than they do in the high mountains. The Mediterranean climate in California produces different fire behavior than the coastal climates of western Washington and Oregon. And no matter how you cut it, the changing climate is amplifying fire risk across the country.

TODAY

Over the last twenty years, wildfires burned bigger and hotter. Fire seasons are now more than two and a half months longer than they were in 1970. We see more destruction from fire as we've built homes deeper and deeper into the forest. Smoke now blankets cities from coast to coast. San Francisco, Portland, Seattle—cities known for clear, crisp Pacific air—are now regularly choked by smoke. And East Coast cities can no longer escape the eerie hue of wildfire smoke in late summer. The fires that were an exception decades ago have become the norm.

Here are some sobering statistics: the average annual acres burned across the U.S. from 2017 to 2021 was about eight million, double what it was thirty years earlier. The federal government spent an average of $2.5 billion fighting wildfires between 2016 and 2020. Thirty years ago, we spent a third of that.

The country's long, complicated relationship with fire got us here. Our addiction to burning fossil fuels for our own convenience created a climate that's out of balance and rapidly changing, causing even more fires. People moved en masse into areas that regularly burn. And for a century, fire suppression has by and large been the name of the game when it comes to fire policy, contributing to crowded, unhealthy forests itching to burn.

Today, we're in what researchers call a fire deficit. That means a gap between how much fire we ought to expect based on climatic conditions and how much fire we're actually experiencing. Montana State paleoecologist Cathy Whitlock says that the historical relationship between temperature and fires is well understood: higher temperatures mean more flames. "We're due for more fires," she says.

At the same time, we're also confronting what some call the "wildfire paradox." Eliminating fire to avoid the damage to forests and property virtually ensures that catastrophic fires will continue to rage. As some in the fire world say, "A fire put out is a fire put off."

Fighting fires takes more of a toll on firefighters than ever. Solutions aren't easy—they require transforming institutions, crafting better policy, changing how we behave, and, perhaps most important, shifting our collective attitude toward fire. The rest of this book is devoted to diving into the complexity and messiness of fire to forge a brighter path ahead. A fiery future is guaranteed. But the extent to which it damages people and communities is up to us.

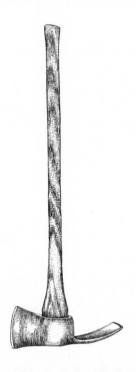

FIGHTING

Across the country, more than thirty thousand wildland firefighters operate to tame the largest conflagrations threatening homes and property. They operate with an almost military level of efficiency and organization, using techniques and strategies based on decades of applied and scientific understanding of fire behavior. This section will help anyone living in a fire-prone area, or with loved ones who do, understand how and why fires near them are fought, and who's on the ground doing the work.

A QUICK HISTORY

Some form of suppressing fire has likely happened as long as people have been living in fire-prone areas. But one place to start the modern, U.S. tradition of wildland fire crews and firefighting is in Yellowstone National Park in the 1880s. There, the U.S. Army was put in charge of putting out fires. Soldiers became the first fire crew to get paid for their service in the park.

Since then, members of the armed services continued to have influence on fighting major wildland fires. During the 1910 Big Burn, for example, thousands of U.S. troops fought the blaze. An all-Black unit of so-called Buffalo Soldiers helped save hundreds of lives in the towns of Avery and Wallace, Idaho. Decades later, during World War II, the government transferred the 555th Infantry Battalion, an African American unit known as the "Triple Nickles," to the West Coast, to learn how to fight fires and defuse bombs that might be sent that way from threats abroad. They didn't put out any fires sent by foreign enemies, but they did fight fires side by side with the Forest Service. By the late 1940s, more than 80 percent of the most elite firefighters were veterans. After Vietnam, too, combat-trained pilots helped pioneer helicopters' presence on wildfires. Even today, it's become common for members of the National Guard to be deployed to fight fires during peak season, and several firefighting crews across the country are made up entirely of veterans.

By the 1930s, technology advances had helped improve the ease and efficiency with which fires could be fought. Felling a tree with a chainsaw takes a fraction of the time it takes to cut a ponderosa with a handsaw or an ax. Getting to the scene in a truck or

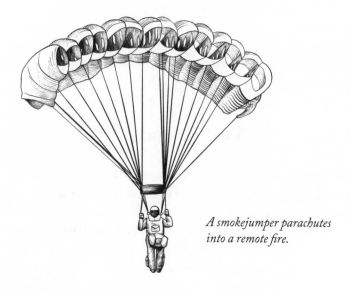

*A smokejumper parachutes
into a remote fire.*

ATV can save hours of exhaustion compared with hiking in on foot or riding on horseback. Soon, a cadre of "smokejumpers" were trained to parachute from planes into remote forest areas to suppress flames. That program still exists today, along with "helitack" crews who rappel into the scenes of fires from helicopters.

As part of the New Deal, President Franklin Delano Roosevelt created what became known as the Civilian Conservation Corps. Over nine years, nearly three million CCC members built thousands of miles of trails, fought fires, and constructed hundreds of "fire lookout towers" in the backcountry where solitary staffers could spot distant flames. The first "hotshot" crews emerged in the 1940s. These are highly trained fire crews who respond to wildfires all over the country. These days, there are over one hundred hotshot crews across the nation, each with about twenty members.

While firefighting itself progressed, research on wildland fire behavior made leaps and bounds. In 1915, five years after the Big Burn, the Forest Service established a research branch. One of its central goals was to better understand, detect, and suppress wildfire. The agency established "experimental" forests, where it could seek to better understand the nature of wildfire. It sought to get a stronger grasp on the weather and climatic conditions that led to big blazes, and to systematize a fire danger rating system so the country could be prepared for hazardous conditions. By World War II, firefighting took on the same ethos as the battles abroad. Fire research, some argued, could also serve to bolster U.S. security from foreign threats.

In the late 1940s, thirteen firefighters died as a wildfire blew up in Montana at Mann Gulch. That tragedy spurred the Forest Service into more research and made it clear that it wasn't just the dollar values of trees that were at stake; it was human lives. In the years to come, the agency established a first-of-its-kind Fire Laboratory in Missoula, Montana. The lab sought to better understand how to prevent fire, predict it, and suppress it. It also set out to understand how using controlled fire could protect forests in the long run. As the century wore on, the Forest Service ramped up its research on fire ecology as well. If firefighting was once a program based mostly on firsthand experience on the fireline, it was quickly becoming a rigorous, scientific operation.

As fires increased in severity and scope into the twenty-first century, so did the cost of putting them out. In the late nineties, federal agencies were spending about $500 million a year putting out fires. Then, those costs began to climb: federal suppression costs averaged about a billion dollars a year as the new millennium began, and would soon explode. Fire suppression began to eat up more than half of the Forest Service's budget. Between 2016 and 2020, average annual spending on wildfires ran about $2.5 billion. In 2021 alone, that price tag was over $4 billion.

The surging costs of wildfire are partly due to worsening fires amid climate change and more people living in fire-prone areas. At the same time, this new era of destructive wildfires has given rise to what some call a "fire-industrial complex." Legions of trucks, tankers, bulldozers, helicopters, and airplanes—often leased from private companies and with a hefty price tag—are components of fighting just about every large wildfire.

Today, tens of thousands of wildland firefighters spend months each year camping and tromping through steep, rugged backcountry terrain in heavy boots, Nomex clothing, and bright helmets. California's firefighting agency, the California Department of Forestry and Fire Protection—or Cal Fire for short—employs about eight thousand people. It's second only to the Forest Service in overall number of firefighters. The country's third-largest firefighter force is a private contractor: a company out of Bozeman, Montana, called Wildfire Defense Systems.

Fire lookout towers have largely been replaced by cameras, planes, weather tracking, and other technology designed to track where fires are occurring. But the story of firefighting isn't a clean arc of forward momentum. Firefighters are quitting en masse due to inadequate pay and health care. They're reporting burnout and mental health issues due to the unprecedented length and demands of fire seasons. They're also struggling with many of the same issues the rest of society is grappling with, including how to incorporate new ideas and more diversity in a field often dominated by a "good ole boys" mentality.

How, where, and why fires are fought can shape how threatening fires become. Understanding how this immense apparatus functions—and where it's going—is crucial to understanding how we might get out of the fiery conundrum we face today. From a practical standpoint, the language used to communicate about fires is specialized but important to understand. Every day when a fire strikes near a community, meteorologists, firefighters, and crew members present what the fire's doing, how firefighters are addressing it, and what's likely to happen next. This section will help anyone affected to understand what's going on.

DETECTING A WILDFIRE

Fires that get caught quickly and early often don't get big and devastating. In many areas of the West, about 95 percent of wildfires are caught relatively quickly and kept to ten acres or less. That's a little bigger than the size of seven football fields. But how do officials detect burns when they're just starting?

That information comes in a number of ways. First, wildfire is often divided into two forms: natural and human-caused. Natural, lightning-caused wildfire was responsible for about 70 percent of the acreage burned from 1992 to 2015 and about 52 percent of the average acreage burned from 2017 to 2021, so it makes up a huge chunk of overall wildfires by land. To keep track of where those strikes might be hitting ground, officials use remote sensing to track lightning and identify general areas where wildfires might sprout up.

Things get trickier when it comes to human-caused burns. Hundreds of millions of people recreate, work, and live in fire-prone places, so there's no remote-sensing technology that can track where someone's likely to start a fire. At the same time, people are responsible for nearly 90 percent of wildfires by number. Those fire starts can come from downed power lines, sparks caused by vehicles, fireworks, smoldering campfires, and so much more. Many fire officials say mitigating the impacts of human-caused fires is an important problem to solve: the season for human-caused fires lasts about three times as long as the season for lightning-caused burns, and fires started by people also travel much faster and threaten more property. In terms of overall loss, human-started fires are responsible for as much as 97 percent of burns that threaten homes.

People often report these fires the old-fashioned way: with phone calls. Somebody might see a flame licking the base of a tree, or smoke in the distance, and call 911. A system of fire lookouts, many of which were built nearly a century ago, also still dot many mountaintops across the West. While some have been

decommissioned or shifted to recreational use, a few hundred fire lookouts are in operation today. It's a far cry from the several thousand that were staffed in the lookouts' heyday, but the workers at the helm spend peak fire season much as the old timers did, with little human contact, scanning distant hillsides for signs of smoke. Artists and poets like Gary Snyder and Jack Kerouac spent fire seasons in lookouts. For many, the experience slowed down the bustle of the modern world. Snyder penned the short poem "Mid-August at Sourdough Mountain Lookout" about his experience:

> *Down valley a smoke haze*
> *Three days heat, after five days rain*
> *Pitch glows on the fir-cones*
> *Across rocks and meadows*
> *Swarms of new flies.*
>
> *I cannot remember things I once read*
> *A few friends, but they are in cities.*
> *Drinking cold snow-water from a tin cup*
> *Looking down for miles*
> *Through high still air.*

Today, there are, of course, many areas fire lookouts *can't* see. So air patrols fly set routes over forests to scout what's happening on the ground from above. Then, there are technological solutions.

Like many other jobs across the country, fire lookouts have been largely mechanized. In many parts of the country 360-degree cameras, infrared technology, and drones all detect fires. NASA even has two types of satellite that can track fires from orbit. Groups continue to actively develop even newer technology that can detect smoke and other signs of wildfire as early as possible.

Our problem isn't detecting fires: it's what comes next—how, where, and when we suppress those fires—that matters.

SUPPRESSION AT WORK

When a fire's been detected, first responders on the scene—often members of local fire crews—"size up" the fire. They assess its location: Who can get there, and when? What's the terrain like? Is it hilly and rugged? Is it a relatively flat grassland? What's the weather doing? What time of year is it and what sort of season has it been? Are gale-force winds forecast for the hours to come? Is it safe? What kind of fuel is around? Could loads of small growth help ladder the flames up into the canopy? What about firefighter safety? Where can the people on the ground access with minimal risk of injury? Officials also assess where the fire could go and the resources they need to get it under control—things like helicopters, tankers full of water, or the numbers of firefighters needed to cut down trees and get rid of other fuel. One firefighting term that plays a huge role in how fires are fought and how many resources get deployed to a burn is *values-at-risk*. That basically means things like homes, structures—all kinds of private property that a fire could burn through.

If decision-makers deem a fire is in need of suppression, "initial attack" begins. That's the stage when firefighters arrive on scene and try to keep the fire from growing any larger. The first, and most simple, tool in firefighters' toolkit is called the *fireline*. That means striking away at the vegetation in a strip of land, creating a barrier devoid of fuel that the flames, in theory, can't cross. Imagine a wide, dirt trail or even a road. If in a remote area, firefighters might dig fireline by hand with tools like the Pulaski and the McLeod, which look like a rake on one end and a wide hoe on the other. Where vehicles can access, they might plow through with *dozers*, or specially designed bulldozers meant for wildland firefighting.

Hose lay often comes next. That basically means hauling hundreds or even thousands of feet of fire hose to the fireline. That hose is connected to a tanker carrying water, or some other water source. The hose helps keep strategic sections of the line wet and therefore less vulnerable to the fire.

In this early phase, the fire usually gets mapped and labeled. InciWeb (inciweb.wildfire.gov) is a resource that provides real-time updates, maps, reports, photos, and more for wildfires across the country. The fire gets a name—often related to where it's located: Kootenai River, Bear Gulch, Williams Lake, Dodge Springs, Big Sandy. If a fire breaks out near you, and you're looking for information, this website is a good bet for checking on the most up-to-date conditions, evacuation orders, fire behavior, and the like.

Fire managers and public information officers often present statistics, based on how crews have contained the burn: a fire is 75 percent contained, or 0 percent contained, and so forth. That refers to how much of the fire's perimeter has been caged in by fireline or other barriers that prevent the spread of the fire. Those could be things like highways or rivers.

Things get complicated when the fire jumps the fireline or burns in terrain where it's impossible to access one end of the burn. That's where understanding another key difference in firefighting tactics comes into play: direct versus indirect attack.

Direct attack means fighting the burn head-on. Firefighters might try to spray water or knock it back with dirt. It also means digging fireline near the edge of the flames. It's hot, hard work—and generally only possible on lower-intensity areas of fires during relatively calm weather.

When things heat up—when the fire's *running*, or moving fast, the wind's blowing, or the fire's just too big—firefighters use *indirect attack*. That means they put some distance between themselves and the fire. Instead of digging line right next to the flames, they might find a natural feature like a road or a trail to work off of—or else dig a line of their own that could function as a more distant fire perimeter. Rather than stop the fire, they try to guide it to a place where it will run out of fuel and contain itself—places like older burns, rivers, or roads.

The goal is to stop or slow the rate of spread of the fast-moving fire. Sometimes, firefighters will light a fire of their own from their control line, meant to burn the vegetation between the line and the oncoming flame front. That's called a *backburn*, *burnout*, or at times, *backfire*.

The game of wildland firefighting is all about manipulating the fire's potential fuel. The main way to do that is by removing as much flammable stuff from the ground as possible. Things get out of control when wind whips enormous flames over the fireline, or embers drift from the flame front to start new *spot fires* away from the central fire.

Some firefighters say direct attack is made nearly impossible in huge megafires, where fires can run miles in a day. Made worse by decades of fire suppression and climate, fires with that kind of extreme behavior are becoming more and more common today.

A fire burns within a bigger context. The country as a whole is experiencing other fires, each soaking up a portion of the overall resources available: things like air tankers, fire engines, and firefighters themselves.

Similar to the "fire danger" ranking you might see at your local Forest Service trailhead, the country also has a broad-scale rating system of how prepared firefighting forces are for the conflagrations they might face. That scale runs from 1 to 5. During preparedness level 1, there are very few fires burning, and practically

WHAT'S IN A NAME?

Superstorm Sandy. Hurricane Hugo. Monkey Pox. Disasters and diseases get names, and the same goes for wildfires. Browsing InciWeb—the database for all the fires happening in the U.S.—can feel a bit like reading avantgarde poetry: the Elmo Fire, the Weasel, Sawtell Peak, Washakie Park, the Potter, Left Fork, Goat Rocks. Who names wildfires, and why do they get the names they do?

When an agency responds to a fire, the first on-site responder or dispatcher decides what to call it. They tend to pick something that's descriptive and immediately useful for firefighters, so they often go with a nearby prominent place-name. Sometimes that's the nearest road, lake, canyon, or mountain. Once named, the fire then has a place in the recordkeeping, decision-making, and communications systems surrounding response.

Sometimes, a single event like a lightning storm can cause lots of fires simultaneously in the same area. Not all of them get their own name. For management efficiency, those groups of fires get merged into what's called a "complex." But complexes can get more...complex. When several fires converge into a single large fire, that's also a complex, as in the 2020 August Complex in Northern California, when the Doe, Tatham, Glade, and Hull Fires converged into a massive million-acre blaze.

Names can repeat, and they're not always all that creative. In fact, during a particularly challenging season in 2015, a fire in Idaho was named "Not Creative."

nobody deployed to fight them. During preparedness level 5, there are dozens of big fires all across the country, and at least 80 percent of the country's fire crews are on the ground suppressing them. At those higher levels of preparedness, decision-makers are faced with tough choices of where to deploy crews.

Resources can be scarce. Those decisions can depend on fire weather forecasts and how many homes and other structures are threatened, and where.

Fires across the country are managed under the Incident Management System (ICS), and each wildfire gets a "type." While crews come and go, usually in two-week hitches, fire operations can last for months, depending on the terrain and vegetation where the fire's burning. As a fire grows or threatens more property, the designation of the fire can change. A type 5 fire is the smallest. Think of a single tree, or an acre burning in a remote area, unlikely to spread much more. A type 1 fire is the largest. Those can involve thousands of firefighters, all swarming in from across the country.

A fire can change "types" as the burn progresses. What might start as a relatively innocuous blaze can kick up with a windstorm or some other weather event. It can skip over the lines built by firefighters to contain it. A type 4 fire could evolve into a type 1 situation in a matter of weeks or even days. As fires approach type 3 and higher, complexity ramps up too. There's not only more firefighters to manage, but there can be local residents and private property at play. Fresh replacement fire crews on the job have to take into account the knowledge that's been accumulated in the weeks before—but also what the fire might do and where to operate as the fire progresses. Those new teams might have to make quick decisions on the ground with little familiarity with the actual area in which they're working.

Even when a forest fire is fully contained, it can still burn for weeks or longer. In that time, extreme weather could blow in and cause the fire to jump the line and grow in size. Or it could just simmer for months, until fire season ends—often through what's called a "season-ending storm."

THE STRESS OF THE JOB

Imagine: it's the height of summer. One hundred degrees. You're wearing heavy Nomex clothes meant to help shield you from the heat and flames. You have kevlar chaps over your pants to protect you from a deadly or disfiguring thrown chainsaw chain. You got four hours of sleep last night. The most in a few days. You don't have cell service and you haven't seen your friends or family in a week. Your whole world is smoke, dirt, and ash.

The days of a wildland firefighter are exhausting—or else, endlessly boring. The day after a firefighter fells trees all day or hacks into the earth to hand-dig fireline, they might spend hours staring at that line, making sure nothing crosses it. Firefighters often go days or weeks with no bed or shower. The weather can be blisteringly hot in the day and freezing at night.

In addition to taking courses that qualify them for the job, they must also pass a "pack test," which requires hiking three miles with a forty-five-pound pack in forty-five minutes. That speed requires a minimum of fifteen-minute miles, or something close to a gentle jog (which doesn't feel so gentle with a heavy backpack). Firefighters generally live in camps while on assignment. Those range from open fields to local fairgrounds to off-season ski hills. They generally work sixteen-hour shifts—or even longer, sometimes through the night. They're on assignment for fourteen days at a time, with only a couple of days off in between fires. The Forest Service recently instituted guidelines that require three rest days for every fourteen days worked. In the field, there's often little or no cell service or connection to the outside world. It's hard to maintain relationships with family and friends, or to meet the obligations of ordinary life, during seasons spent on the fireline.

Firefighter pay varies widely based on experience and employer. An entry-level federal wildland firefighter may earn about $25,000 for a six-month assignment, including overtime pay. But that can vary greatly. Cal Fire employees, for example, can

Essential gear wildland firefighters carry with them while on assignment.

earn as much as two times what a Forest Service firefighter with comparable experience makes. Private firefighters may earn much less than someone on a federal crew—or, depending on the position, more.

It takes a unique sort of person to become a firefighter. Someone who loves the outdoors, who tolerates or thrives in suffering, who can stand the disconnection to the outside world, and who loves the strong community and culture of firefighters around them.

The job takes a physical, mental, and emotional toll. While wildland firefighters have dug line around conflagrations for more than a century, scientists have only just begun to assess the unique demands of the job.

Firefighters have long experienced what's known as "camp crud"—or a hacking cough on assignments—but studies looking at the effects of smoke inhalation on crews have only begun over the last decade or so. While the science is still in its infancy and long-term effects are hard to isolate, one study shows that firefighters with a ten-year career have a nearly 25 percent higher chance of lung cancer and heart disease than the rest of the population. Looking at folks with a two-decade career, those numbers jump to 25 to 39 percent. Other studies show risk roughly in the same ballpark.

Brent Ruby, a physiologist at the University of Montana, wanted to look at the physical toll the act of fighting fire takes on the body. Ruby had studied nutrition among elite athletes, but he reasoned that firefighters needed to know more about how much energy they were taking in and using. If they were burning a lot more calories than they were consuming, for example, that could take a tremendous toll on the body and mind—and end with more death and injury on the fireline, where the stakes of the job are so high.

He figured that the only way to get the science done was to spend time at fire camps—traveling the country as what he described as "a Johnny Appleseed wildland firefighter physiologist evangelist of sorts"—and analyzing what happens when crew members are actually getting the work done. That entailed things like body weight measurements, carefully controlling and monitoring food intake, and taking samples of urine, blood, and saliva. He also used naturally occurring isotopes called *tracers* to find out how much oxygen the body consumes in a given day and how much carbon dioxide it produces. Putting that ratio together, he could compute energy use on the job.

He discovered that firefighters burn about four thousand to seven thousand calories a day—about two to three times more than the average person. That sheer energy output shows that firefighting is hard and unforgiving work. Eating that much also takes time. Time most firefighters don't have, so they often run a caloric deficit, the cumulative effects of which can be hard on the body. You might expect firefighters to get stronger as the season goes on, but Ruby observed some of them actually getting weaker. He says that taken along with the other issues firefighters face—lack of sleep, choking down smoke—"all those stresses pooled together create this sort of composite overarching stress."

Knowing where the stress points lie—and how to overcome them, whether it's through mitigating smoke hazard or planning food intake appropriately—can help contribute to better decisions and performance on the ground. "Knowledge is freakin' power and knowing what the demands are and knowing how a body heats up on the fireline and why it does is power that saves lives," Ruby says. Given all the stresses and demands of the job, his work suggests that when it comes to confronting the risks and rigors of firefighting, "fitness can be the ultimate counter-measure."

BURNOUT

The history of firefighting is full of tragedy. In 1949, thirteen firefighters were killed when a fire blew up and spread over more than three thousand acres in ten minutes. A brush fire in California's Mendocino National Forest swept over and killed fifteen firefighters in 1953. Fourteen firefighters were overrun by the flames of Colorado's South Canyon Fire in 1994. The Yarnell Hill Fire in Arizona killed nineteen members of a hotshot crew in 2013.

Wildland firefighters die at a rate six times higher than structure firefighters. From 1990 to 2019, more than five hundred wildland firefighters were killed on the job. Heart attacks, car wrecks, and aircraft accidents caused the overwhelming number of those accidents. Statistically, it's more dangerous, for example, driving to and from a fire than actually digging line or conducting a burnout. But still, firefighters face danger and unpredictability every day on the job.

After every major incident, agencies hold reviews to figure out what went wrong and how they might prevent it from happening again. "Firefighters understand and embrace that inherent risk," Dan Cottrell, an incident commander and longtime smokejumper, says. "We try to minimize it and mitigate it, but there is inherent danger in what we do and there always will be."

In his time on the ground, Cottrell has seen the industry of firefighting change substantially. Cottrell says that decades ago, fire

ED PULASKI

Ed Pulaski helmed a fire crew during the Big Burn of 1910, and his leadership during the fire made him the stuff of firefighting legend.

Outside Wallace, Idaho, as the forest seemingly exploded, all hope seemed lost. But Pulaski had an idea. He knew of an old mine tunnel, not too far away. He gathered his forty-five men and set out for the shelter. They made it just in time, as the fire overtook their trail. The men were panicking, crying, and praying. Pulaski stood at the tunnel's entrance, hanging wet blankets and dousing the flames with water scooped with his hat. His men passed out. Soon, Pulaski did too.

The next morning, as the surviving crew members stirred, they awoke to an unconscious Pulaski. He remembered hearing someone say, "Come outside, boys, the boss is dead." Pulaski stood up. "Like hell he is," he shot back.

Six men on Pulaski's crew passed away, but he saved dozens of lives. Today, his story is part of the cultural fabric of wildland firefighting. And his presence touches firefighters' lives in other ways too. Pulaski grew fascinated with a very particular tool. It was two-sided: one end looked like an ax, for chopping, the other like a hoe. The tool had likely originally been devised for planting trees, not stopping fires. But Pulaski saw its potential. He made improvements to its design, and today, the Pulaski is still a tool you'll find in use on just about any wildfire across the country.

was a seasonal event. The Forest Service responded accordingly: "They built the firefighting apparatus around a seasonal workforce," he says. Now, fire season begins in March in the Southwest and extends well into the winter in other parts of the country. Studies show that fire season is nearly three and a half months longer than it was a few decades ago. Working so hard, for so long, firefighters run the risk of burning out. Part of it is the physical demands of the job: the miles-in, miles-out hikes with heavy packs take a physical toll. Same goes for digging line and hauling tree limbs around, and the smoke inhalation.

But emotional aspects of the job are just as tough or tougher. Firefighters spend most of their work lives away from their friends and families. When they return, they barely have time to catch up on the obligations of adult life—much less make time for their loved ones. In the field, they're often out of cell service range. At the same time, nearly every longtime firefighter knows someone who's been critically hurt or died on the job. When fire season ends, it's easy to feel out of touch, dislocated—almost like a soldier returning from war.

Longer seasons add to the toll, and Cottrell says the firefighting workforce hasn't kept up with the changing fire conditions. He says these changes aren't the Forest Service's fault. The world is changing faster than the apparatus of government can. He says individual "fire seasons" aren't really a thing anymore. It's a lot closer to a fire *year*. "I worry about those guys that have been out on assignment starting in June and they're still out in December," Cottrell says. "I worry about whether they can personally handle that stress load and then recover."

While the science on the physical demands of the job is reasonably developed, formal studies on what the job means for mental health haven't yet been born. Nelda St. Claire has made it her life's work to better understand what wildland firefighters are facing and how to help them.

St. Claire grew up just outside Yellowstone National Park in a firefighting family. As a youngster, she'd hear chatter on her father's radio. There'd be talk of burnovers, when a fire runs and overtakes firefighters on the ground, and air tanker crashes. It was clearly a dangerous job. But growing up so close to it, something about it grabbed her.

She says she became a firefighter for the same reason lots of young folks make wild decisions: her dad told her not to. When she started, she loved the seasonal lifestyle: fire in the summer, skiing or traveling to Mexico in the wintertime. "It just stuck," she says.

As she spent time digging line and working on fires across the country, she too experienced some of the trauma she'd heard on the radio when she was young. She witnessed burnovers, had people close to her die. Then, in the early 2000s, she was in an event where the fire burned so fast and so intensely, she had to deploy the lightweight shelter she was carrying to survive. Waiting it out in that tiny, claustrophobic shelter is a terrifying experience.

After St. Claire's shelter deployment, her crew received something called CISM, or Critical Incident Stress Management. It's a crisis response meant to address the mental toll that traumatic events can take. St. Claire says that the fire world is a culture unto itself. And anybody dealing with the mental health needs of firefighters needs to be *of that world*. But the people sent in to provide *her* with CISM services about two decades ago didn't know anything about the culture of firefighters. "Their lack of knowledge of the culture was offensive," she says. "I think it did more harm than good. In fact, it was more psychologically traumatizing than the incident itself."

St. Claire says one aspect of fire culture is *toughness*: the ability to cope with anything that gets thrown at you. To shrug it off. To barely acknowledge it. Working long hours digging line, toughness is an asset. But she says that trait can mean shrugging off mental issues, just like physical ones. It can amplify the already existing stigma of talking openly about this stuff.

"We didn't talk about it," St. Claire says of the stress. "Somebody might think you're crazy or weak or that you couldn't do the job. You didn't talk about it."

St. Claire's first experience with CISM planted a seed that would blossom in the years to come. St. Claire herself started working full-time in crisis intervention in 2015. For four years—until she retired—she served as the National Critical Incident Stress Management Coordinator for the Bureau of Land Management and the National Interagency Fire Center in Boise. NIFC is the nucleus of fire response across the country. There, she saw how all the federal agencies responded to firefighter trauma from the highest level. And she wanted that response to have more of a cultural connection to the people on the ground.

To make that happen, she needed to learn a little more. In those four decades working for the feds, St. Claire learned the value of data. She started keeping track of all the CISM requests for firefighters. It was an under-the-table tally that she kept to herself. And eventually, she started keeping track of one number in particular: suicides among wildland firefighters. It had long been assumed that firefighters took their own lives at a much higher rate than the rest of the population. But there were no numbers to back up what the community was experiencing.

She kept a separate database, validated by coroner's reports or family members, where she could track suicide. She noted how long they'd been fighting fire, potential causal factors, and how they passed away. This is not an official agency statistic. This is something that St. Claire kept track of alone, because she understood its importance. The numbers are unofficial and incomplete, because she counted only suicides she could confirm. The ones that got reported.

The story her numbers told was dramatic. There was a huge spike, she says, from about 2010 to 2018. Numbers went as high as thirty in 2015, and twenty-two the year after. Those numbers are conservatively ten times higher than in the general population. At a certain point, the number of firefighters dying by suicide surpassed the number of firefighters passing away through accidents on the job.

Barely any academic work has focused specifically on suicide among wildland firefighters. But one study presented at a conference surveyed more than twenty-six hundred current and former firefighters. The survey found a 17 percent rate of depression, a nearly 14 percent rate of PTSD, and about a 20 percent rate of suicidal thoughts in the last year. Those numbers are two to ten times higher than the rest of the public.

"When we start to see mental health problems and post-traumatic stress-related illnesses, it's often not tied to a single incident, it's tied to a series of incidents," St. Claire says. "Now, there really isn't an off season, so the stress becomes cumulative."

Instead of those numbers continuing to balloon, though, something very different has happened. "We have seen an increase in calls to crisis call centers—a huge increase," St. Claire says. "And a decrease in suicide."

In 2019, she counted five suicides. The year before, even fewer. But—more people calling in to address their troubles. It's important to point out: these numbers don't say that this story is over. It could be a fluke. Things could change in the years to come. What it does say is that maybe firefighters, as a whole, are getting a little more open. Talking more freely. Feeling less isolated, more understood. "Talking helps," St. Claire says. "I learned that with my own traumas, talking helps."

Members of the firefighting community are starting to go public with their own stories. Firefighter Lucyanna Labadie wrote a blog post after choosing to leave the field.

She mentioned that on her last assignment, her crew replaced a hotshot crew on which one member had been killed by a falling tree. "Questions, comments, concerns?" her captain asked in a briefing. Nobody spoke up. She worked that assignment in a daze. She did her job—but, she wrote, "it occurred to me that I was suppressing more than just fire." She left the Forest Service when she got home. She processed her emotions for a month and gave them voice in her post.

"Why do we fight lightning start fires in the middle of nowhere?" she wrote. "Why do we attempt to stamp out every last fire? Why do we see fire suppression as heroic? Why do [we] suppress our pain when our brothers fall?...Why do we push down grief, anger, hurt?"

Bré Orcasitas, a firefighter for more than a decade and a half, says it's common to feel like you don't have the ability to voice discomfort or vulnerability, for someone to say, "No, I'm not okay." There's a part of firefighter culture that says, "you knew what you were getting into when you signed up for this job." But that's not true. When most folks start in their early twenties, the firefighting world is mysterious, adventurous, and enticing in its challenges. "Nobody pulled you aside in your training in the beginning to say 'Hey, just so you know, if you stick with this job, it's almost impossible that you're not going to end up knowing at least one person who's died from doing this job. You might be catastrophically injured.' That's not part of it."

But, Orcasitas says, firefighting culture is changing fast. Those dynamics affect training, and even demeanor on the ground—like during briefings or digging line after something bad happens. She says things like practicing mindfulness and meditation are more acceptable too. "I think a big component of it is there are so many people in the fire community who've been affected by some traumatic incident," Orcasitas says. "It's like you can't even pretend at this point that it's not happening or that people are okay."

She's also seen a slew of firefighters leaving the ranks—folks like Mike West. He's a seventeen-year veteran firefighter who posted a public letter of resignation in the summer of 2020. It got a lot of attention online.

"In my career, I was almost burned over four times," West wrote. "I came within a few feet of being killed by a falling tree on two occasions. I've had a few close calls with vehicles, rolling objects, and a large angry bear." In the letter, he said the Forest Service has gotten pretty good at attending to physical injuries in the field. However, the biggest threat to his life was the symptoms of PTSD. He said the training he received dealt with injuries during work, but not with recognizing and preventing mental health distress.

He encouraged firefighters reading the letter not to be embarrassed about the mental and emotional struggles they may be having. He said firefighters *are* getting more open. But there are still *too many* who stay quiet. "I hate to say this," he wrote. "But I'm more concerned I'll lose a friend to a mental health issue than to another fireline fatality." He wrote that he hopes the letter will resonate with even one person, somebody like himself—who might feel less alone to know others are going through the same things. And despite his struggles with PTSD, he was clear that he had a positive experience overall in the Forest Service, full of laughter and learning. Stuff he'll carry with him in his next career move, as he steps into the world of teaching.

Even Brent Ruby, the physiologist, is trying to help outsiders better understand what firefighters go through. He wrote a kids' book, *Wrango and Banjo on the Fireline*, about two dogs who fight fire in Montana. The book is meant to be a source of inspiration and wonder and help kids learn about things like navigation and maps and Pulaskis, and to feel inspired about the natural world. But he also hoped to open a door into the internal lives of firefighters.

"Maybe it's idealistic or I don't know how to explain it, but I thought, gosh, firefighters are just these wonderful characters. But they have chinks in their armor like every other mom or dad or aunt or uncle or grandparent or whatever," he says. "And they have difficulty communicating with the littles in their lives like we all do."

HOW TO SPEAK THE LINGO

Wildland firefighting is full of jargon that can be tough to decipher (a Forest Service glossary of wildfire terms runs nearly two hundred pages), but critical for nearby residents to know for getting a grasp of what's happening on the ground. Understanding how fires behave, what containing them actually means, and the complexities of a wildland fire operation can help people affected by burns breathe a little easier in the smoky air.

Firefighting teams look at fuel, weather, and topography, or the layout of the land they approach. They can manipulate the fuel; the rest informs the safety and strategy of the operation. So the key to wildland firefighting is eliminating fuel, or the stuff that burns.

Fireline a line of ground without vegetation that will presumably stop or direct a fire's progress. Firefighters dig line by hand, using Pulaskis, McCleods, and other tools. Bulldozers and other heavy equipment are also used to make larger firelines.

Backburn, backfire, and burnout terms for intentionally putting fire on the ground and burning vegetation against an active flame front to deprive it of fuel.

Indirect attack a method of fighting a fast-moving or high-intensity wildfire that uses existing fuel breaks to stop a fire's advance. A control line is staged at an existing feature, like a road, or firebreak, lake, or river, often far from the fire's front.

Direct attack any action that directly engages a fire. This can be digging fireline, spraying water, or dropping retardant from the air.

Black the area that's already burnt. A common saying in wildland firefighting is "keep one foot in the black." That means, always have an already burned area you can escape into, since that likely won't burn again.

Green fuel-laden area that hasn't yet burned.

Anchor point a strategic location from which to start building fireline. The goal of an anchor point is to prevent the fire from burning around the fireline, pinning fire-fighters from behind.

Control line any constructed or natural barrier used to impede a fire's progress.

LCES (lookout, communi-cation, escape routes, safety zones) safety essentials of any wildfire operation. Lookouts are scouts, positioned to watch both impending fires and fire crews, to warn of any potential disas-ters. Communication refers to how information is conveyed to crews on the ground—this could be from radio to word of mouth. Escape routes are how firefighters will get to safety, should the situa-tion become unsafe on the ground. Safety zones are areas where fire-fighters can find respite from a dangerous situation. All elements of LCES may shift in just a matter of hours as fires change behavior.

Values at risk the things threat-ened by a wildfire. Often, this refers to homes and private prop-erty. But values at risk can also include water supply, power grids, and cultural or historic areas that a wildfire might destroy or alter.

Size up the process of quickly and accurately assessing a fire and its critical characteristics. This should include a name for the fire, its location and size, an assessment of the terrain, values at risk, weather conditions, the fire's behavior, what resources are needed, and what caused the fire (if known).

Mop up cleaning up an area that has burned. This can include tend-ing to dangerous trees, removing remaining fuel that could reig-nite, trenching logs so they don't roll, and anything else to make the scene as safe as possible.

Incident Action Plan (IAP) a plan of action covering twelve to twenty-four hours, developed by the incident commander assigned to a wildfire. IAPs specify the tac-tics and support activities planned for a certain time period.

Incident commander the head honcho on a fire and one of as many as dozens of people manag-ing the complex planning, safety, strategy, and operations on a conflagration.

Hand crew general-purpose wildland firefighters. Hand crews are typically eighteen to twenty people and work on digging line, clearing trees and brush with chainsaws, and setting controlled burns with drip torches. The Forest Service specifies three crew Types, referring to the experience and capabilities of the group. A Type 1 crew is a team of Hotshots. A Type 2 IA crew is less experienced than a team of Hotshots, but can break up into 4-6 person teams to engage in initial attack (IA) with the fire. Type 2 crews are more entry level. They do not engage in initial attack, but dig line and engage in other critical work less close to the fire.

Engine crew a team of up to ten firefighters attached to a fire engine, tasked with initial, direct engagement of a wildfire. They use a variety of tools, primarily relying on hoses and water.

Hotshots an intensively trained team of firefighters primarily tasked with directly engaging a fire and digging hand lines.

Helitack crew a team of firefighters trained and certified in the use of helicopters to fight fires. Helitack crews are transported into fires to directly engage the flames and are often the first people responding to the incident.

Dozer a heavy machine with steel tracks and a front-end blade. Designed for clearing trees, brush, logs, and other materials down to the mineral soil.

Air tanker a fixed-wing aircraft designed or retrofitted to transport and deploy fire retardant directly to fire or adjacent land. The Forest Service uses a variety of aircraft with capacities ranging from eight hundred to eight thousand gallons.

Tender any kind of ground vehicle tasked with supporting a critical wildland firefighting asset. Tenders can deliver fuel and repairs to dozers, refill helicopters or planes with fuel, or deliver water to engines.

Engine a truck or other ground vehicle that can transport and pump water, via hoses, onto a fire.

Fine fuels fast-drying, highly flammable materials such as grass, leaves, or pine needles, less than a quarter inch in diameter.

Helibucket a bucket that hangs from a helicopter by a cable and is used to transport and apply water or retardant directly onto a fire. Helibuckets can be dropped into a lake or river to refill with water and can hold as much as twenty-six hundred gallons. That water can be dropped directly onto a fire.

Parallel attack a fire suppression tactic in which a line is dug approximately parallel to the fire's front, but some distance away. Once the parallel line is set, firefighters set fire to the ground between the line and the fire, robbing the fire of fuel and stopping its advance.

Swamper and faller a faller cuts down trees using an ax or chainsaw. This role requires skill and experience with tools and knowledge of tree characteristics and behavior. A swamper supports the faller by clearing brush, trimming limbs, shuttling supplies such as fuel, and watching for danger.

Dirty August a firefighter term for a month when fires in the West tend to increase in size, severity, and complexity.

Nomex a brand of flame- and heat-resistant material worn by firefighters.

Head imagine a fire like a charging bull. The head of the fire is the horns, where the fire is stomping ahead at full force.

Flank the sides of the fire, to the left and the right of the head.

Rear the part of the fire opposite of the head. To continue the analogy above, the rear of the fire is the bull's kicking hind hooves.

Rate of spread the speed at which a fire is moving from its origin point, usually affected by wind, moisture, and slope.

Chain a unit of measurement, commonly used in wildland fires. Eighty chains equal one mile, so one chain is sixty-six feet long.

Fire fingers long, narrow sections of fire extending out from the fire's primary front.

Pockets unburned areas between the fire's fingers and the main front of flames.

Island an area of unburned land within the perimeter of a fire. Wildfires don't always torch everything in their path.

Spot fire an occurrence when embers drift from the main fire and settle on vegetation or other flammable material, igniting new flames.

Flashy fuel light, highly flammable materials such as grasses, shrubs, leaves, or needles. These fuels are often the cause of home ignitions.

Torching when one or more trees go up in flames.

Ladder fuels flammable materials, like low limbs and small, young trees, that can allow a fire to move from the forest floor up into the canopy, increasing the intensity and potential growth of a fire.

Widow-maker a safety hazard for firefighters in which a branch or treetop is loose from the main tree. When that tree is cut down, the loose limb may fall off away from the main tree, potentially onto the firefighter below.

Snag a dead, often fire-killed, standing tree that presents a hazard for firefighters on the ground.

Slopover when a fire spreads outside the perimeter of a control line.

Ground, surface, and crown fires a ground fire burns mostly below the surface of the earth, on roots and in duff on the forest floor. A surface fire burns pine needles, shrubs, and small trees on the forest floor. Here, flames might be immediately visible. When those fires grow larger and "ladder" up trees, they can ignite crown fires. That's when the canopies of trees light with flame. These are often the largest, hardest to contain, and fastest-moving fires. When one canopy ignites, that fire can begin jumping from tree to tree.

Fire triangle heat, oxygen, and fuel. Fires need all three components to burn. Firefighting works by depriving fires of their fuel.

FIRE RETARDANT

When a wildfire strikes near a community, news stories often show planes dropping mists of bright red on forests. The image is nearly apocalyptic, an otherworldly shower descending on the landscape. The substance raining down from the sky is fire retardant, meant to slow the spread of fire and keep new swaths of forest or grassland from igniting. The federal government spends millions a year on the stuff—but how effective is it?

The active ingredient in these bright showers of fire retardant is ammonium phosphate, a common fertilizer. The mixture was first developed by the corporate chemical giant Monsanto in the 1960s, but today's specific mixture—mostly what's known as PHOS-CHEK—usually comes from Missouri-based Perimeter Solutions. A number of studies have tested ammonium phosphate's ability to suppress fire, and the chemical is even used in paints and other materials as a fire retardant. In the case of PHOS-CHEK, the ammonium phosphate is mixed with water, some chemicals that prevent it from corroding the aircraft, gumming agents that make the mist stick to trees and other vegetation, and red dye so that pilots and firefighters can see where it's landing and what still needs to be covered. Primarily, it gets deployed on fires in order to lower their intensity, slow their advance, keep more fire from burning, and even "mopping up" or helping make sure a fire simmers out. Retardant is also used to strengthen existing fuel breaks, like roads or open meadows. The Forest Service uses this stuff a lot: in 2021, the amount of fire retardant deployed could fill more than thirty Olympic swimming pools.

The Forest Service insists that retardant drops are an important tool in the firefighting kit, but they admit they don't know what the substance actually means for success fighting fire on the ground. Since there are so many factors that go into fire growth—from weather to fuel type to the tactics used to fight it—it's hard to pin down the exact role of fire retardant in the efforts. The nonprofit Forest Service Employees for Environmental Ethics filed a lawsuit in 2022 to limit the Forest Service's use of retardant because of its harmful effects on fish and other aquatic species. It was the third time the group sued the Forest Service over its use of retardant. The lawsuit alleges that the substance violates the Clean Water Act and potentially the Endangered Species Act. That's because when fire retardant enters rivers, creeks, and streams, it becomes ammonia, which is highly toxic to fish. The Forest Service's own draft Environmental Impact Statement on the use of retardant says that the substance "may affect" fifty-seven threatened and endangered aquatic species and is "likely to adversely affect" more than thirty additional aquatic species. At the same time, FSEEE claims that on the most extreme fires, where retardant is likely to be used, it isn't much more effective than simple water drops. The problem isn't a small one either. Between 2012 and 2019, according to government data, the federal government dropped more than 100 million gallons of retardant on fires. About 760,000 gallons of that was dropped on or near rivers, lakes, and streams. Over that same period, the Forest Service spent about $58 million to $100 million per year on retardant drops.

So why does the federal government continue to use the stuff? The Forest Service itself refuses to comment on ongoing litigation. But there are several possible answers. The first is just that the substance works. It's one tool in the toolkit, and we need all the help we can get combating the megafires we see today. But there might be other drivers too. Some call the use of retardant "CNN drops," since they make such great fodder for news footage and give the public the appearance that agencies are doing everything in their power to put out the flames. On the ground, they say the drops mean little for the behavior of the fire. Politicians often blame agencies for not doing enough when fires blow up, so putting on a show with retardant drops could simply be part of the public relations necessary to show the public due diligence. Part of it could also be agency culture overall. "When it's fighting a fire, it's in a war," Andy Stahl, executive director of FSEEE, told NBC Montana. "That's the way the

Forest Service mentality views fighting fire. It's a war, and collateral damage is just a part of business as usual."

Stahl says that collateral damage isn't just for fish and wildlife. In 2020, for example, fifteen people died fighting wildfires. Nine of those deaths were aviation-related.

WHO ARE ALL THESE FIREFIGHTERS?

Tens of thousands of people work as wildland firefighters in the U.S. In fact, in September 2020, the peak of a particularly big fire year, almost thirty-three thousand firefighters were deployed across the country. As wildfires threaten more communities across the country, that number is growing. And at the same time, the makeup of that firefighting force is changing too.

In 2021, the federal government employed more than eighteen thousand firefighters. About half of that workforce was in the Forest Service. The remainder was assigned to other government agencies—like the Bureau of Land Management and the National Park Service. That still leaves more than fifteen thousand firefighters working for other, nonfederal agencies. State departments of forestry have firefighting forces of their own. Cal Fire is the largest in the nation, employing around eight thousand people. Other, smaller jurisdictions have their own forces too—including counties, towns, and other rural areas. Plus, municipal fire departments get involved when fires threaten neighborhoods.

One unexpected source of firefighting labor in the U.S. comes from prisons. Since World War II, California, for example, has used inmates as wildland firefighters through its Conservation Camp program. Other states, like Nevada, Arizona, Colorado, Montana, Wyoming, and New Mexico, have similar programs. The statistics indicate incarcerated firefighters suffer worse outcomes than their professional peers—more injuries, death, and long-term health effects. Some consider this exploitative, but some inmates cite the experience as critical to their rehabilitation.

As wildfire seasons have gotten worse and federal spending has ramped up over the last couple of decades, an industry has developed around firefighting, and that's including the rise of private fire crews. The third-largest employer of firefighters in the nation—behind the Forest Service and Cal Fire—is Wildfire Defense Systems, based in Bozeman, Montana. WDS contracts with insurers to protect homes both before and during fires and deploys firefighters across a twenty-state region. The company says its work with insurers doesn't come with any extra charge to policyholders. WDS is just one example, though. The National Wildfire Suppression Association represents over three hundred private firefighting firms. Some of these firms occasionally contract out to wealthy homeowners to provide additional fire-security measures for their homes. Famously, Kim Kardashian gushed over the private crew she and Kanye West hired in 2016 to successfully protect their $60-million mansion from the Woolsey Fire. But much of this work often comes from these private firms contracting their resources and crews out to the federal government to respond to incidents across the country. So the people on the ground during big burns aren't just federal and state employees; they're private contractors too.

Some claim that the multibillion-dollar business of fighting wildfires has given rise to a "wildfire industrial complex." With a nearly open checkbook, injecting that kind of cash into wildfire can change the bottom line of where, why, and how wildfires are fought. When private industry gets involved, for example, there might be more profit-based incentive to suppress fires even when they're good for the ecosystem and don't threaten any communities. There are other issues with the business of private fire crews too. Some crew members feel like they don't earn the respect of the federal and state employees on burns. They claim that they're often treated like pieces of machinery for rent, not like people.

At the same time, firefighting is an intensely white and male profession. The numbers change year to year, but in 2021, 84 percent of federal wildland firefighters identified as male. One 2020 survey of Forest Service employees in the Pacific Northwest found that about three-quarters of female firefighters felt like they didn't belong because of their gender. Seventy-two percent of employees surveyed said that they were white. Congress has held hearings over discrimination and harassment in the firefighting force. The Forest Service and other federal agencies insist that they are actively trying to change the composition and culture of the firefighting force. After allegations of misconduct, the Forest Service launched a review of its practices and formed an action plan to overhaul its reporting processes. Along with other agencies, the Forest Service currently offers women-specific trainings and internships that can provide mentorship and community. In 2020, it also established an office dedicated to employee well-being and addressing the root causes of cultural barriers in the service. It's an ongoing effort with no easy answers.

As climate-driven fires continue to escalate throughout the year, threatening more communities, the country's firefighting forces will undoubtedly continue to change as well.

FIREFIGHTER PROFILE: LILY CLARKE

Lily Clarke grew up in Montana's Swan Valley, a small community sandwiched between the Mission and Swan Mountains, next to the Bob Marshall Wilderness. In 2003 the Crazy Horse Fire blew up nearby. It burned about eleven thousand acres in all and threatened homes there. Firefighters flooded her community of about three hundred people. She says that the fire was billowing smoke so dense she couldn't play outside. She remembers trying to sit and watercolor but could barely see the paper she was painting on. Her family packed up their belongings to get ready for evacuation. They put their sheep and horses on a friend's property that was less at risk. It was her first real experience of what fire could mean for people—and the business of fighting it.

In her community, there were constant debates over how to manage the landscape: where and how to log, and what that could mean for future burns. The threat of fire was something that divided people. But then, when the fire had simmered out, out of the soil sprouted something that brought people together: morels. Morels are mushrooms with caps that look like delicate, oblong honeycombs. Fried up, dried out, put on pizza or in an egg scramble—they taste delicious. In that part of Montana, foragers flock to recent burns the spring after fires to collect them. In the Swan Valley, the same people who might have

been arguing in town halls were suddenly side by side in the forests, gathering mushrooms. What once had divided now united.

Later, during Clarke's first year of college, another fire hit the area. This time, it was about a mile and a half from her home. Her family hadn't prepared their property for the possibility of flames. "It was a real reckoning with knowing that I lived in a fire-prone area," she says. She remembers a conversation with her mother, who said that they'd find a way to move forward if a fire took their home. They'd prefer to lose property rather than put the lives of firefighters in danger.

In college, she studied morels—the postfire mushroom she'd seen bring her community together—and expanded her studies globally when she got a Fulbright fellowship in Nepal. When she got back from her fellowship, she joined a fire crew near her hometown. "I was given the gift of not only being a firefighter but being able to work in the landscapes and with the people that I had grown up with," she says. At the same time, she kept pursuing her education, earning a master's degree in systems ecology, with a focus on wildfire. The two paths—firefighting and ecology—informed each other: she wanted to experience how fire decisions are made on the ground, and she also loved the experience of forming an intimate relationship with a wild, natural process.

Clarke served on a federal engine crew. That means she worked for the Forest Service and her crew traveled with an "engine," a truck designed to aid firefighting efforts by holding and pumping water. No two shifts were exactly the same, and she described a spectrum of emotions when on a fire: boredom, exhaustion, excitement, fear, and awe: "You could spend fourteen days sitting on a line watching the green side of the fire to make sure that no embers cross over there and never really do much. You could be digging line around a massive fire for fourteen days and feel broken after two of those. You could be bringing your engine out and spraying down the fireline or during night shifts, burning off the fireline."

At first glance, it might seem like their mission is straight from Smokey Bear: put out fires. But wildland firefighters don't just suppress fires. They help protect property. They put fire on the ground to burn away vegetation in a careful, controlled way. Still, as an ecologist, Clarke confronted the tension between putting out fires and letting them burn. "I absolutely struggle with that," Clarke says. "I absolutely struggle with seeing decisions made to put out wildfires in areas that have been adapted to fire, that can be far away from communities and should be let burn in those areas. But I'm not a firefighter to tell. I'm here to learn."

Interacting with fires big and small on the ground, she sees firsthand the thorny and complex decision-making that goes into managing burns all over the country. If decision-makers decide to take it easy on a fire, or to let it burn, for example, and that fire erupts and threatens a community, "you will be given direct responsibility for that fire going that way." She doesn't necessarily mean the legal responsibility—though in some situations that could be the case. She means there's a real emotional and moral burden that decision-makers shoulder if things go awry. Also, when fires are "let burn" and get out of control, politicians nearly always pipe up with new calls to suppress all blazes.

Through her background in ecology, Clarke sees that society as a whole needs to reshape its relationship with fire. She says that people need to develop social resilience—to stop living in fire-prone areas and to take appropriate measures to reduce their risk when they do. And they also need to develop ecological resilience—to allow wildfire a place on the landscape where nature needs it. Embedded in the reality of fighting fire, Clarke remains optimistic that society can reshape its relationship with flames. She says she's seen legions of young firefighters, so many of whom are understanding the deeper, ecological role of fire in our landscapes.

MEGAFIRES AND GIGAFIRES

In 2017, the longtime firefighter and writer Bill Gabbert wrote an article about a massive fire in British Columbia called the Elephant Hill Fire. It had grown to nearly two hundred thousand acres.

The article was on Wildfire Today, a website dedicated to providing nitty-gritty details about fires across North America. Most of the writers are also firefighters—Gabbert had worked in that world for over three decades. He'd seen fires getting bigger and bigger. And by that time, "megafire" had become a well-known term for wildfires over one hundred thousand acres.

Curious about the ever-intensifying fires across the country, Gabbert posted a comment on his own article: "If we call fires that exceed 100,000 acres megafires, what would we call 1,000,000-acre fires?"

A user on the site with the handle "kevin9" offered a simple answer: "Gigafires, of course." The term caught on—and thus *gigafire* was born. The term has since made headlines across the country.

Neither gigafires nor megafires were necessarily new. The Big Burn, for example, blasted through more than 3 million acres. The Yellowstone fires of 1988 burnt more than 1.6 million. Gigafires continue to be relatively rare, although the August Complex in California did burn more than a million acres in 2020. However, megafires continue to threaten forests and communities across the West. As of 2022, more than two hundred megafires have scorched the landscape since 2000. Fifteen burned more than 500,000 acres.

MONEYBALL FOR FIRE

When a fire goes big, things get complicated. The people in charge need to figure out where the fire might go, what it might threaten, where to dig line, what natural features could blockade its spread, what the wind and weather might do, and much, much more. And all that's happening when other fires are going on across the same region, or across the country. The stakes are high—a blowup could threaten a community, a watershed, or firefighters themselves. Bad decisions could mean resources don't go where they're most needed. Now, data and modeling are helping to inform how the most important on-the-ground decisions are made in an increasingly complex environment.

"I think that folks would be very surprised to know how thoughtful and methodical and how much science is behind the management of fire and the suppression of wildfires," Rick Stratton, a fire behavior analyst with the Forest Service, says. "There's so much more going on behind the scenes." And behind that curtain, the way those decisions are made is changing, fast.

The data revolution in firefighting arguably began with Forest Service researcher Mark Finney's FARSITE in the late 1990s, a model that turned the world of fire behavior prediction on its head. The modeling system looks at the topography of the land, fuels, wind, weather, and predicts how a fire might spread across the landscape. Modeling before that was an abacus. After FARSITE, it became a calculator. As time went on, analysis got more complex—and began to offer probabilistic models of how the fire might spread. Essentially, the projection calculates thousands of different scenarios to figure likelihood of fire spread, and its interaction with the landscape and weather as conditions might change on the ground. These aren't public models, but they help fire managers decide the probable course of a fire and what might be at risk.

As computer power increased, so did agencies' understanding of the landscape. Many more acronyms and models and names got in the mix. Most important, agencies could begin to input and map tons of variables: vegetation and fuels on the ground, historical fires, where people live, sites with a lot of valuable resources like drinking water, spots that could be dangerous for firefighters, and even natural barriers on the landscape like roads, rivers, and ridges. From a technical standpoint, all this information is revolutionary. All of it, interacting together, creates fine-grained portraits of entire landscapes. It now informs how and where fires are fought.

The modeling was first applied to fires on the ground in 2016 at a meeting with lots of fire folks in the Pacific Northwest. When researchers presented the potential of all the analytics, someone stood up with a smile on his face and said, "Oh my gosh, this is just like *Moneyball* for fire!" The movie, which dramatizes how the data revolution reshaped baseball, had just come out. Stratton remembers the moment well—the comparison, he says, is a little corny. But it's also accurate. "Essentially we're using analytics to help in wildfire management instead of using analytics to help win baseball games."

The world of analytics, like so much of wildland firefighting, is full of acronyms. One to be familiar with is PODs, or potential operational delineations. Pioneered by a research team from the Forest Service's Rocky Mountain Research Station, Colorado State University, and Oregon State University, PODs look at conditions and places where fires stopped burning in the past or charged right through. They look at fuel changes, breaks in fuel, and more—and create bite-sized chunks of the landscape in which fire can easily and safely be controlled. Covered in PODs, the landscape looks like a quilt or a turtle's shell. The edges of each quilt square reflect natural points that could serve as barriers to fire spread.

Within the patchwork of polygons, the models also show how a forest could change if it burns. In many areas, that value shows up bright red on the map; a fire there would burn hot and huge. But other areas show up green on the maps. After decades of suppression, those areas *need* fire. If a fire comes through, it would likely be a healthy burn and help restore the landscape. In essence, these analytics help fire managers engage a fire before it even starts. "The term that we use is 'Select the right ridge, not the next ridge,'" says Tim Sexton, a program manager for a Forest Service wildfire management research and development team. He means that finding the most effective and strategic natural features can be crucial when confronting fires. Finding the "right ridge" can make or break efforts to contain a fire. "And the analytics we've got help us select the right ridge to put the fireline on."

PODs are challenging. They extend beyond just Forest Service land. That means if, say, Forest Service officials on the ground want to plan ahead of time to be ready for fire season, they might need to get local landowners on board to use their property as part of the effort. But with enough coordinating and planning ahead of time, the models can help maximize the efficiency of how firefighting resources get distributed. Dave Calkin, a supervisory research forester at the Rocky Mountain Research Station, says that every fire season, something like a "common pool resource problem" emerges. Basically, "everybody's grabbing at the same pie and people in first grab as much as they can." However, even though folks on the ground are incentivized to use as much as they can on their own, individual fire, fire resources as a whole are limited. There are only so many fire crews, engines, planes, and helicopters. The models could help inform exactly how many suppression resources are appropriate for a fire and exactly where they ought to go. Importantly, they can also help inform decisions about which fires need to be fought in the first place.

"We saw how analytics just transformed sports. And we're saying the analytics need to transform fire management," Calkin says. The challenge, it turned out, was getting these analytics accepted and used on the ground.

PUSHBACK

In a scene from *Moneyball*, a group of scouts sit in a room discussing what players would be ideal for the team. Two scouts start throwing out ideas that make little sense to the others. But the math shows that those out-of-the-box choices would generate the stats they need to succeed. Everyone else, of course, pushes back. "Baseball isn't just numbers," one scout says. "They don't have our experience and they don't have our intuition."

Calkin says that scene holds up pretty well in firefighting too. "I've had fifteen years' experience of bringing analytics to fire managers," he says. A handful of them jumped right into using them. "But then there's the majority that are like, what the hell do you know? You're not one of us."

For decades, the strategies used to fight fire have been decided by folks working for local districts of the Forest Service, incident management teams, and the like. With years of experience, those decision-makers have observed how fires behave and how suppression tactics work. The people at the top of the chain, in theory, earned their way through their experience. That—not the education, work in other fields, or new skill sets they might bring—got them to the point where they were entrusted to make the tough decisions it takes to manage wildfire. They've seen fires simmer and seen them explode. They've had to make tough, life-and-death calls. And they've done it well. The science writer Malcolm Gladwell popularized an idea in behavioral psychology: that it takes ten thousand hours to become an expert at something. These folks have ten thousand hours under their belts, and then some.

The ten thousand hours idea is compelling. Anybody can be an expert if they just put in the work. In terms of fire management, it makes clear who should be making decisions and why. But that idea, it turns out, is a bit of an oversimplification. Crucially, the Israeli psychologists Daniel Kahneman and Amos Tversky (academics later called them the "Lennon and McCartney of social science") had this other really important idea: human brains are bad at making decisions when confronted with complex and uncertain scenarios. No matter how much experience someone has, the brain contains unavoidable biases; it's far from a rational, calculating machine. Gut feelings often tend to push decisions in the *wrong* way.

This plays out in all kinds of ways. In their work, Kahneman and Tversky identified all kinds of fallacies, misperceptions, and biases that just about everyone relies on in daily life. For example, they identified what they dubbed the "representativeness heuristic." That's a group of errors based on a simple fact of the human brain: we compile our past experience into prototypes to help make sense of patterns we intuit. Those prototypes, though, are based on our limited experience—not probability or statistics. Kahneman writes of a classic example. Imagine an acquaintance is describing to you someone you've never met:

> Steve is very shy and withdrawn, invariably helpful but with little interest in people or in the world of reality. A meek and tidy soul, he has a need for order and structure, and a passion for detail. Is Steve more likely to be a librarian or a farmer?

By and large, folks responding to that question say "librarian." Meekness and tidiness more closely align with the stereotype of that profession: tweedy, glasses, organized. However, statistically the answer is resoundingly "farmer." There are far more farmers in the U.S. than librarians. The human brain, Kahneman and Tversky argue, saves itself energy by making shortcuts like this—and it's often wrong. This is a simple example of this sort of bias, but it applies just as well in wildfire, especially since climate change, fuel loading in forests, and development into the WUI is creating wildfire behavior that people on the ground have never experienced before.

The researchers behind PODs and the analytics revolution in wildfire argue that using these maps, numbers, data, and modeling

FIRE ON THE PLAINS

When you think of wildfire, large trees—often pines—sending flames and smoke high into the air come to mind. But grass fires are every bit as common and just as dangerous.

Carl Seielstad, a professor of fire science and management at the University of Montana, says that grass fires start fast, grow fast, and end fast. A fire in a forest ecosystem, by contrast, might simmer for weeks or months before cool, wet weather snuffs it out. One reason for the difference is fuels. Grasses are "flashy fuels"—meaning thin, flammable material.

For firefighters, grass fires present new challenges and a few advantages. There are a lot of firefighting strategies that are just more feasible in a grassy landscape that's more open and has more fuel breaks like roads and bodies of water. In addition, the fuel type is more homogeneous (and therefore predictable) compared with a mountain ecosystem, where there's rugged terrain and radical changes in things like tree type and elevation that can affect how things burn.

Grass fires are every bit as important to ecosystems as forest fires. Historically, grass fires would have ripped across huge swaths of the country. The Lewis and Clark expedition documented continuous fire from almost North Dakota to the Rocky Mountain Front, flames forty, fifty, eighty feet tall, ripping across the landscape. "It must have been a jaw-dropping sight," Seielstad says. "Fires were vast, beyond comprehension, beyond anything we see today."

could help eliminate the errors that individual decision-makers can create when it comes to wildfire management. That could mean fighting fires more efficiently and making communities safer from burns—and also easing up the response on fires that don't pose as much of a threat.

Rick Stratton is working to make the data and models more digestible to people using them on fires. First, he created an online dashboard with maps and layers containing fuel, PODs, fire risk, and more. He's trying to make analytics accessible to the people who might be hesitant. He likens the uptake of the technology to giving out samples at a grocery store. Initially, he says, you might get just a few people coming to the table and asking for more. Others might take a little bite and walk away or avoid the table entirely. "But with time," he says, "you're starting to see more and more people coming to the table. You're starting to see people open to the idea."

Stratton says the data and models aren't meant to supplant the gut decisions and intuition of incident commanders and other decision-makers on the ground. Rather, the analytics and the lived experience can complement and inform each other. Experience *is* important, especially in wildfire. Someone with years of experience on fires might notice subtle changes in topography or weather that inform fire behavior—things a less-experienced firefighter might not even perceive. A key to success with PODs—those patchwork polygons of control lines covering a landscape—is engaging with folks who know the terrain well, to accumulate a wealth of micro-knowledge that can enhance those maps.

At the same time, bigger-picture changes are making the need for analytics more tangible. Climate change and the legacy of fire suppression are bringing fire intensity and behavior that are creating unprecedented situations. Tim Sexton has worked in wildfire since the 1970s. He's been on fire crews, led fire crews, worked as a ranger, and directed multiple Incident Management Teams. Still, he says, "We're experiencing situations now that are not in individuals' experience in the past." In other words, given the changes we're seeing, experience isn't enough.

The team's analytics were used on 8 fires in 2016 and 11 the year after. By 2021, that number had increased more than tenfold, as they were used on at least 113 wildland fire incidents. At least 40 percent of those 113 utilized PODs in their efforts. Today, more than sixty national forests have engaged with PODs to help plan out the best strategies before the fires even begin. Stratton says that the Forest Service is hiring more staff to help get these analytics out of offices and into the field. "It's become integrated into our wildfire management system," he says. "So that's really exciting to see. And it's fun to see people see the light."

MORE FIRE ON THE GROUND

PODs and wildfire analytics can help maximize the efficiency of suppression resources. But there's another, much-bigger-picture benefit too. They could help reshape what fires we let burn and set the stage for lighter, safer, and healthier fire seasons in the future.

To understand how analytics might do that, you first need to understand an outlier in fire management: the Forest Service calls them "resource benefit fires." Other people call them "managed fires" or "proactive fire use." They function very much like prescribed fire, or flames intentionally put in areas to restore their ecological function. But in this case, managers let actual wildfires burn. But they don't just step away and let them go; they carefully monitor them. And they might be our last, best hope of confronting the wildfire crisis.

The scale of the wildfire problem is enormous. Millions upon millions of acres of land—both public and private—are overgrown and ready to burn. Right now, fire managers successfully squash well over 90 percent of fires when they're tiny. Those remaining fires are, by and large, burns that get away from those initial attacks. However, many folks in the fire game think that strategy needs rethinking. "Mother Nature can be our worst enemy and can be our best partner at times," Stratton says. "The fires that we shouldn't put out, we're putting out . . . and the fires we want to put out, we can't."

Sexton was assigned to fires in the Shasta-Trinity National Forest in Northern California in the late 1980s. For the most part, he says he was working to suppress small, lightning-started fires. Then, suddenly, he watched as a single fire burned over the fifty or sixty smaller fires he'd played a part in suppressing that season. It struck him: "Maybe we need to rethink what we're doing with our overly aggressive fire response."

Decades later, that attitude had solidified. "We can't keep fire from the forest and have a healthy forest," he says. "The risk associated with fire increases each year that you remove fire from a frequent fire system." After all, many forests in the West historically experienced low-intensity fire as often as every ten years, or even less. In 2011, Sexton was working as a district ranger, managing part of the Boundary Waters, a massive wilderness area partially in northeastern Minnesota. A lightning-strike fire hit, and he saw the potential. The area hadn't seen fire in more than a century. He took a chance and made the call to manage the fire for "resource benefit"—in other words, to carefully let it burn safely. His call wasn't made with the assistance of the PODs and other cutting-edge models available today, but this was a rare opportunity, he thought, where the fire could burn safely and restore the ecosystem.

Suddenly, though, the weather changed and the fire blew up. Instead of the twenty-five-thousand-acre burn he'd expected, it was over ninety thousand acres. Sexton says he took lots of heat from the public and the press. It threatened the lives of some hikers and Forest Service crew members. But in the end, no one was hurt or killed—though it remains the largest fire on record in the area's history. The fire provided a sixteen-mile-long buffer that could prevent future conflagrations. It also restored habitat for the species that call the area home.

Looking beyond the Boundary Waters, the forests of the West would have historically been composed of a mishmash of old fire scars. Some areas might have burnt within the last one or two years, others a few decades ago, still others more than a century ago. Each scar might stop another, new fire in its tracks. Older burns, over time, filled back in with new growth—providing the conditions necessary for new, healthy burns going forward. In this sense, historical fires were self-regulating. But our success at fire suppression has put an end to many of those boundaries. Sexton asks: "Are we just putting out those fires that are good fires at a small size? And then the bad ones are the ones that we struggle mightily against and spend all of our resources at and have all the injuries and fatalities on?" He says that if we allowed more fire to burn, something novel would happen. A more sizable patchwork of burned areas would emerge. The boundaries of those burns could in turn limit the size and severity of the big fires that plague us today.

Here's a recent example of how that might work. In May 2017, lightning struck a tree in the Pinal Mountains of southern Arizona. Not far away sat the community of Globe—population about seven thousand. Months before the smoldering began, officials in the area had begun planning for the next burn. The local fire management officer saw a problem: fire starts were common in the arid region—and were mostly always managed with full suppression in mind, aimed to keep them as small as possible and put them out as quickly as firefighters could. As a result, the area around Globe was primed to burn, and if things went haywire, a nearby fire could easily reach town.

The local fire management officer approached the mayor, city council, and other stakeholders armed with one more tool on the belt: PODs. The officer wanted to get local buy-in and pitched the idea: if a fire starts in the right area, with the right conditions, let's try something different. Instead of hitting it hard with everything we've got, let's let it burn a little longer and in a larger area. PODs provided a network of boundaries that could help keep it in control.

When the Pinal Fire started, the team in charge knew exactly where the control locations would be; they'd planned for this event. They put fire on the ground to burn back toward the lightning-started flames and reduce surface fuel. The operation went relatively smoothly, at least for a while. On the sixteenth day of the fire, things changed dramatically. "It got sketchy," Calkin says. Humidity dropped, temperature jumped, and winds skyrocketed. Large wildfires get their own meteorologists to analyze the weather and what changing conditions might mean for fire behavior. But beyond a few days out, there's always a layer of uncertainty, even with the best weather models. A couple of weeks after a decision is made to let a fire burn, for example, the weather could change dramatically. In this case, that's exactly what happened. Unexpected high winds blasted through the chaparral, sending up flames as high as fifty feet. The fire made aggressive runs uphill—the sort of intense fire behavior that the management team was hoping to avoid. They had to call in a type 1 Incident Command team to prevent the worst-case scenario.

In the end, the community of Globe remained safe. And within another week, the fire was under control. All told, it burnt more than seven thousand acres. The fire wasn't quite a resource benefit fire, left to do its natural thing. It also wasn't a full suppression fire. It was somewhere in between. The fire was allowed to burn a relatively large area. It wasn't managed to stay as tiny as possible. Things got dicey, and tactics had to change, but that initial decision to let the fire be a fire would prove to make all the difference in the long run. From an ecological perspective, the fire had reversed a decades-long trend. And from a safety standpoint, the fire made a huge difference. It became a buffer surrounding the community of Globe.

Fast-forward to 2021. Another fire hit the area. This time, it was of a very different character. Called the Telegraph Fire, it was a megafire burning with extreme intensity. It seemed unstoppable, and it was heading right for Globe. At its height, it burnt more than 180,000 acres. However, the Pinal Fire scar stood in between the Telegraph Fire and the community of Globe—and firefighting teams leveraged that to their advantage. They used that fire scar as a fuel break to keep the community protected. And it worked.

"You can see that Pinal Fire protected well, flat out, no doubt about it," Calkin says.

However, he was clear the story wasn't one-dimensional. He says he was telling the story of the interaction between the Pinal and the Telegraph Fires at the University of California, Berkeley, recently, and one attendee said, "Hey, I support that." He was a firefighter, and he'd been there when the fire blew up. "I thought it was a great fire, but I almost got killed on that fire. I've never been as afraid in my life as when the chaparral started burning."

Every fire, even if it's managed well, has a level of risk. Calkin recognizes that. But he says thinking differently about what fires we can tolerate also requires reformulating how we think about that risk. "A five-thousand-acre fire, you might be able to have some level of control," he says. "That fifty-thousand-acre fire or the one-hundred-thousand-acre fire, you got no shot." The challenges arise because in most communities, the pressure nearly always comes from the same direction: put the fire out, don't let it burn. There's little tolerance for fire on the landscape. Fire managers get rewarded for preventing fire, not for letting it happen. A full suppression strategy usually gets nearly 100 percent public support, Sexton says. When a prescribed fire or a "managed fire" gets out of control and threatens communities, politicians and locals blame fire managers for inaction and laziness.

Using analytics to let fires burn more acres has the potential to restore ecosystems and reduce future risk. Other ideas for getting more fire on the landscape include introducing more prescribed fire (read more about that on page 134), but those projects require environmental review, which can take years, staff, and funding. The data revolution in fire could help the country reach the scale it needs to create resilient forests and communities. The road to getting there won't be easy. It will require reformulating our collective relationship with risk.

Continuing business as usual, stamping out nearly every fire on the landscape as quick as we can, just sets us up for even more devastating fires down the road. But getting more fire on the landscape means building tolerance for smoky skies and even the occasional escaped burn. For the Forest Service and other agencies, that means more communication with the public, more collaboration with communities, and more transparency about where and why fires are fought. "That's the big challenge," Calkin says. "How do we accept short-term risk to avoid bigger, longer-term risk, because we're not hardwired to do that."

HOW TO HELP FIREFIGHTERS

Homes present special challenges for firefighters. In a residential neighborhood, the stakes are high, and so is the complexity. When working a fire in the forest, firefighters can move around and adjust their attack to the behavior of the blaze. But a home sits in a fixed location, which can be tougher and more dangerous to defend. There, you—as a homeowner—can help set the stage to make sure firefighters are safe.

Firefighter Bré Orcasitas described a recent trip she took to the San Diego area. She says her profession provides a certain lens. Looking around anywhere, she sees fire risk. In this case, she saw sagebrush-covered hillsides, topped with homes. It seemed like a single spark could send fire ripping uphill. "I felt uncomfortable even being in Southern California," she says. "It's just like, any minute something's going to pop off." The kicker there is even when it burns, that brush grows back the next year. That fire risk doesn't go away. These are places where fires will always have to be suppressed. And still, we keep building in areas ripe for continuous flame. "Over and over again, people are putting themselves in harm's way."

Orcasitas says the best way to protect firefighters is to make your home more fire resilient. That means doing basic renovations like replacing wood roofs, relandscaping yards to reduce fuel, and making sure that your home doesn't have any vulnerabilities to floating embers. (For more on that, see the ADAPTING section.)

She says that firefighters will drive right past homes that are likely to burn. That means homes with tightly spaced trees and shrubs in their yard, surrounded by wood mulch, maybe going right around their wooden, flammable deck, are simply too dangerous to defend. "It's cold, hard facts," she says. "You just drive past and you go to the places that you can save that have done some things to protect their own homes."

At the same time, firefighters say homeowners need to shift their expectations. People don't expect federal agencies to fight tornadoes or hurricanes. Instead, they focus on preparation ahead of time and recovery after the fact. So why do they expect firefighters to throw so much at controlling wildfires? "Should we be asking these firefighters to be walking a hairy edge of risk where at any minute everything could go really, really bad?" one firefighter, Rod Moraga, says.

MANAGING

Living in a fiery world requires grooming the landscape, or managing the stuff that burns—namely, trees, grasses, and other fuels. That can come in the form of cutting, trimming, pruning—even careful, controlled burning. How can we manage a hotter world, with an increasingly flammable landscape?

FOREST MANAGEMENT

Imagine a forest.

What do you see?

A leafy or needle-layered canopy, sunlight dappling through. The sound of footsteps on soil and decaying needles. There might be mushrooms growing beneath the thick brown of trunks. A deer or an elk trotting by. Trees have seized our collective imagination for millennia. Forests can be dark and brooding, mysterious—but awe-inspiring in that incomprehensible mystery. Or they're beautiful nearly beyond belief, a refuge from the buzzing and hectic modern world.

Anywhere you look, there is wonder in trees. They help produce the oxygen we breathe and sequester potentially planet-warming carbon from the air. Some twisted, gnarled bristlecone pines in the Sierra Nevada are close to five thousand years old. Some of those trees might have lived through the Trojan War, much of the rise and fall of ancient Egypt, and the construction of Stonehenge. Trees, like people, live in communities. Aspen entwine roots and share nutrients. Some trees form alliances of sorts with other species to survive hardship. Scientists have found birch trees sharing sugars with firs. Fungus grows from roots like fine white silken thread. These threads, called mycorrhizal networks, allow trees to "communicate" with one another and transfer nutrients through this underground "wood wide web." Trees can emit substances to make their limbs taste bad to browsing animals, and they can produce pheromones to attract wasps to attack caterpillars hard at work munching on leaves.

But for as long as forests have entranced us, they've also provided the raw materials that have built our world. The *Epic of Gilgamesh*, one of the oldest written works ever discovered, tells of deforestation around modern-day Lebanon. Ancient Greeks and Romans logged—paving the way to the deforestation of much of the Mediterranean. Indigenous people across North America cut down trees for millennia, even using fire to fell them. Deforestation in the U.S. ramped up as the country expanded west. The Forest Service, founded in the early twentieth century, came to be with an explicit goal of working against the unlimited exploitation of the country's forests and conserving timber for future generations.

For decades, foresters have called the way we harvest trees from forests "management." Managing trees and shrubs and forests might seem like a counterintuitive concept. Forests can seem wild, beautiful, beyond human influence. But at the same time people have been influencing forest ecosystems for nearly the entirety of our existence as a species. Of course, the control we exert over those landscapes has ramped up in recent decades. Even a "hands-off" approach, choosing not to intervene in how a forest grows and matures, is a form of management. In his 1989 book *The End of Nature*, Bill McKibben writes that there's no landscape on the planet that's escaped the reach of people's impact. The world we live in is one of our creation—both its beauty and its destructiveness.

At the same time, how we treat or "manage" our forests is critical to the wildfire problem. Part of the issue today is what many say has been a century of forest *mis*-management: the putting out of all wildfires, leading to an unnatural, unhealthy—and highly combustible—growth in certain ecosystems across the West. Many call for allowing fire to play its natural role on the landscape when it doesn't threaten people and properties. Foresters and scientists also say introducing fire of our own making—ramping up our use of "prescribed fire"—is crucial to restoring forest health and reducing wildfire risk in the future.

Letting fires burn and introducing more of it are two of the tools in the "forest management" toolbox. There are many other tools, too, including reducing small-diameter trees in forests near communities, promoting trees of different age classes in a forest system, and straight-up logging. One way to think of forest management is: we can manipulate the ecosystem to some extent, or fires will do so in a much larger, more merciless, and impactful way. The U.S. Forest Service is, by and large, the arbiter of what kind of work will get done and where it will happen.

From the time of its creation, the bottom line of the Forest Service was timber. Starting with Gifford Pinchot, federal foresters managed forests with the idea of "sustained yield," meaning the perpetual production of lumber, and the economic benefit that comes with it. Trees were treated a lot like crops. As Teddy Roosevelt once wrote, "The object of our forest policy is not to preserve forests because they are beautiful...or because they are refuges for the wild creatures of the wilderness...but the making of prosperous homes."

Trees, after all, are a crucial piece of our built environment as well as our natural one.

Wood products in the U.S. make up about 4 percent of the overall economy. But forests are tied to so much more: wood builds homes and buildings, tables and signs, paper and packaging;

it helps us heat. Forests are also essential to agriculture and water supply, and they're where increasing numbers of people recreate.

In the 1960s, the Forest Service began to recognize that growing and cutting down trees was only one of multiple uses for which it ought to manage forests. But many still saw timber production as its primary function, with board feet produced as its bottom line. By the 1990s, as logging projects were bogged down in lawsuits and the timber industry across the Northwest declined, broader ideas about humanity's role in the environment shifted, and forestry began a profound change. It wasn't just timber or recreation or agriculture, in theory, that the Forest Service ought to manage for, agency officials declared. It was the well-being and functioning of the entire forest *ecosystem*—that bigger natural system in which everything else is contained.

In the decades since, ideas about the broader ecosystem, the importance of wildfire, and the science of forest health have been incorporated into how the federal government manages forests. But in practice, there's still no clear public consensus on how forest projects should look. "Management" can mean protecting from exploitation, promoting resilience, or finding measures of what experts call "forest health." It can also mean aggressively cutting down trees to promote a vision of forestry that prioritizes economic interests over ecological ones. Forest management is ultimately about values.

Familiarity with the suite of strategies used to manage forests—and what they mean for fires—is critical for people living in or near fire-prone areas who want to understand the activities going on in their backyards. The federal government has a robust system of public input and environmental review for forest initiatives, and understanding how these projects work can help you get your voice heard in the future of forest and fire management.

FIRE ECOLOGY: WHAT IS "HEALTHY"?

Just outside the college town of Lawrence, Kansas, the university has set aside a small tract of "experimental" land. There, trails wind their way through rolling hills of plains and woods. The area is an *ecotone*, or a transition area between two different habitats and plant communities. In this case, oak and hickory forest meets tallgrass prairie. This is a tiny little parcel of what once covered vast tracts of the U.S.—only about 4 percent of the historical tallgrass prairie remains. But nearby, an enemy is encroaching. Nonnative red cedars are thriving, turning the plains into woods and gulping down water on which prairie plants depend to survive. The decline of the prairie and the thriving of trees like red cedars is all thanks to fire. Or rather, a lack of it.

Kansas isn't a place people think of for its hills or its forests. But look in the right spot, and it has both. It's also not a place associated with wildfire. But the history of flames there, too, is rich and deep—and crucial to the ecosystem. The landscape burns fast and it burns often—historically, fire came through tallgrass prairie every one to five years. The sudden violence of flames can feel menacing. But fire was a friend to this ecosystem, like so many others. Fire kept the prairie healthy. It made soil more fertile and eliminated saplings of trees like red cedars. Keeping out fire set the stage for the prairie to dwindle away. As settlers developed the area, they created breaks in the vast expanses of fuel that stopped flames. The hooves of livestock and the wheels of wagons, and later, automobiles, trucks, and trains, spread seeds that wouldn't normally grow here. For someone who's grown up here, it's all they've ever known. Walking those trails, even the nonnative red cedars might feel pristine and natural. But—is this area "healthy"?

Zooming out, the example of this single tallgrass prairie in Kansas has a lot to say about the state of our forests and ecosystems all across the country. In areas like this one, a dearth of fire has led to red cedars encroaching on prairie. Elsewhere, in seas of sagebrush, pinyons and junipers are having a similar effect. In ponderosa-dominated western forests, many areas are choked out with small-growth sapling ponderosas and Douglas fir. That creates more fuel for fires, and all that growth can "ladder" flames up to the canopies, where fires can be more intense and destructive. Forest floors are littered with duff and pine needles that haven't burnt in decades. The lushness we imagine as natural and characteristic of wild spaces is often brought about by our own practices: namely, excluding fire.

Some scholars say "health" isn't the right term at all for how we think of forests. More important is the idea of "resilience." A *resilient* forest can withstand the threats of growing human populations, recreation pressure, wildfire, and a changing climate. Paul Hessburg, a Forest Service research ecologist and University of Washington affiliate professor, studies forest resilience in the Northwest. In some of his studies, he's compared historical photos of forests to what's there today. In most cases, the forests and hillsides of the early 1900s were much more open than the exact same areas are today. They had meadows and windows to the earth below the canopies, where today there is nearly continuous forest. That historical landscape was maintained by fire. Without it, forests filled in. The landscapes we've grown up with our whole lives in many cases aren't "natural" or "healthy" at all.

In many parts of the country, scholars say, we're experiencing a fire deficit. That means there's a serious lack of fire based on what we'd expect given the climatic conditions. Historically, millions more acres burnt every year than we tolerate today. That's primarily due to human activity—putting out fires and fragmenting the landscape as we develop. As that deficit grows, the likelihood of enormous, devastating fires grows too. That deficit must close at some point. It's not if, but when.

Forests are changing, and fast. Some of that is due to the changing climate. Other ways that's happening are due to us. We kept fire off the landscape. We continue to develop and build deeper into the wildlands. Now, when extreme-intensity fires burn through, they might burn everything, making it hard for trees to regenerate. They can lead to extreme erosion and cause entire hillsides to degrade. Or they might scorch the soil *too* hot, making it partially sterile for years to come. In many areas, overgrown, "unhealthy" forests are partly to blame for the situation we're in today.

One argument goes: many of the reasons that forests are in an unnatural, "unhealthy" state come from human intervention. That could be from roads, deforestation, climate change, and often, a century of fire suppression. By eliminating natural forces that self-regulate our forests, they've become out of whack. And many foresters and scientists say it will take our intervention to keep them resilient. In that "experimental" plot of land in Kansas, for example, one of the "experiments" involves bringing fire back to the landscape through prescribed burning—intentional, low-intensity fire put on the ground under very specific conditions. In the same way that medications might get prescribed to you for illness, land managers prescribe fire for parts of the country's prairies and forests that need it.

Federal and state agencies also propose logging and "fuels reduction" projects that could make forests healthier—or impoverish them, depending on how you think about it. Federal land managers want to up the pace and scale at which that work gets done. And all this is contentious stuff. People don't want smoke in the sky that comes with prescribed burning—or the risk of those flames running away. Others worry that logging can change the character of forests or lead to a decline in wildlife habitat. Thinking about how you conceive of forest health is crucial as forest managers across the country balance competing values—from timber to recreation to aesthetics.

Ultimately, what forests across the West and the country look like is up to *us*.

FIRE-ADAPTED SPECIES

Plants and animals have been living with wildfire since, well, since there's been plants and animals. Some plants, like the giant sequoia and ponderosa pine, have thick bark that can withstand intense wildfires. Other plants actually need fire to survive. Jack pine, for example, have what are called "serotinous cones," meaning they're tightly closed and covered with a thick resin. When a fire comes through, the resin gluing the cone shut melts, releasing seeds and allowing new trees to eventually sprout. Chaparral—plants like manzanita and other common shrubs in California and Oregon—have seeds that can lie dormant in the soil for years until activated by the heat of a wildfire. Ceanothus plants, like buckbrush and California lilac, even encourage fire through a flammable resin that coats their leaves. Other species grow new appendages after burns. Trees like quaking aspen and bigleaf maple resprout from belowground after fires burn through.

Taken together, these adaptations show the symbiotic relationships that develop between fire and many species. But the country's effort to keep natural fire off the landscape has severed the lifeforce required for these fire-dependent species to thrive.

INDIGENOUS FOREST MANAGEMENT:
A CULTURE OF BURNING

Human hands have altered forests for millennia.

Before the country was colonized, Indigenous people used fire to promote growth and reduce pests and for harvesting food. They used it for war and signals; to clear ground for hunting, gathering, and travel; for materials with which to make baskets; for spiritual and religious purposes. Tribes all over the country had a deep and sophisticated relationship with the landscape. That relationship neither started nor ended with fire. Fire was one process among many that informed soil health, hydrology, wild game and biodiversity dynamics, fungi, and more.

The fires intentionally started to groom the natural world weren't isolated. They shaped broad sections of the continent. Studies in the northern Great Plains, the Pacific Northwest, the redwood forests of Northern California, the oak and chestnut forests of the Appalachians, ponderosa pine in the Rockies, the tallgrass prairie of the Great Plains, the soaring plateaus of Utah, and the Jemez Mountains farther south—and many more areas—all show the ecological impact of Indigenous burning before white settlement.

This practice, known as Indigenous burning or "cultural burning," continued until white settlers had colonized the West. Those settlers saw fire as an enemy—not as a friend, a tool, or even a "gift," as had the people living on the land before them. Fire had shaped the landscape that those settlers gaped at in awe: the forests and plains, the mountains and valleys. "It's ironic that the landscapes so appreciated by the early explorers and colonists actually were created by the very fires they feared and disliked," says the ethnobotanist Nancy Turner.

Germaine White, a member of the Confederated Salish and Kootenai Tribes (CSKT), spent decades working for the tribes in cultural preservation. She began to understand the historical role of wildfire on her tribe's land when working on a project about the traditional names of places. She realized that the natural attributes of many landscapes—with names like "Big Meadows"—had grown out of sync with the natural world. Meadows and forests had filled in and grown clogged. Much of the traditional knowledge of specific sites was dislocated from their present state. And much of that dislocation was thanks to the exclusion of fire.

White's study of the traditional knowledge of her tribe led her to fire—and then to the intentional effort to suppress and erase that knowledge. In a newspaper account she found from the *Missoula Pioneer* in 1875, for example, she read that a group of Pend d'Oreille people encountered "officers of the international line" near the Canadian border while they were hunting. The group was setting fire to the prairie grass to maintain an ash layer that would make up a fertile seed bed for the next year. Two of them were shot and killed by the officers for setting fire. "A pretty powerful disincentive to continue your culture where you're shot and killed," White says.

The U.S. government's war on fire grew entangled with its war on Native culture and peoples. The murder of tribal members by government officers is one example of how that war was waged. As the U.S. government forced Indigenous people off their land and systematically slaughtered bison—on which many tribes depended for food and economy—it began a legal, educational, and frankly militaristic campaign to erase fire from the landscape. In 1850, in what would become California, a law known as the Act for the Government and Protection of Indians made intentionally setting fires illegal. In Karuk Ancestral Territory, on what today is the Klamath National Forest, one district ranger suggested leveraging the trust a well-known missionary had garnered in local communities to discourage the use of fire. She would travel up and down the river, the ranger said, talking to families. "This woman can do more in one season towards causing the Indians to adopt our theories in regards to fire than we can do in five." The terms often turned violent too. Elsewhere on Karuk land, a different district ranger wrote about the "Indian incendiary problem," which he understood in the racist thinking of the time as the result of "the drunken Indian." He contemplated the best punishment for setting fires. "One sheriff says a rubber hose filled with buckshot is perhaps the best," he wrote.

The elimination of fire was bound up with efforts to "Kill the Indian, and save the man," as one U.S. Army officer, Richard Pratt, put it. Pratt helped found the first off-reservation boarding schools, in which Indigenous children were forcibly assimilated into white culture. As the federal government coerced Native children from their homes, it also stripped away the size of reservation land through legislation known as the General Allotment Act. This act essentially subdivided reservation land. Instead of communal territory, it allotted acreage to individuals and families. Then, if the amount of reservation land was higher than the amount of land given to tribal members, the federal government purchased that surplus and sold it off to non-Indian individuals and companies. In this way, tribes across the country lost about ninety million acres of reservation land between 1887 and 1934. The federal government and the timber industry tried to make sure the same bottom line dominated in forests, whether on tribal or federal land: the more logs, the better. On the Flathead Reservation in Montana, nearly half a billion board feet of timber were removed between 1917 and 1928. Some of those ponderosas were so old and enormous that individual trunks filled entire railroad cars. Indigenous cultural elimination and white assimilation were all tied up together. Eliminating the use of fire was one way among many that traditional knowledge was severed from its practitioners.

By the twenty-first century, a great deal of traditional knowledge around fire had been lost. But it wasn't gone forever. Tribes across the country are bringing back cultural burning—not an easy mission. "It's like walking out of the room, turning the light switch off, coming back two hundred years later and expecting to find everything working when you flip the light switch back on," White says. Some tribes, like the CSKT, have taken forest management

Ladder fuels, like low-hanging branches, shrubs, and small trees, transport fire from the forest floor into the canopy. There, in the crown of the forest, the fire burns faster and with more ferocity.

of their land away from the federal Bureau of Indian Affairs and into their own hands. In the case of her tribe, the new forest plan sought to replace timber production as the driving force on the land. Instead, they'd try to manage for pre-European forest conditions, reverse the impacts of fire exclusion, and seek a balance between the human and ecological needs of the landscape.

White says tribes are now finally being included in conversations about forest management. In a 2010 workshop with seven tribal elders and about twenty scientists and resource managers, both Native and non-Native, elders noted that their grandfathers had complained "that it was ridiculous for white people to stop Indian burning and to suppress natural fires." They also described an irony of what's at play today: white colonists put a stop to their burning, and now, white people are coming to them to ask what can be done and how to restore the landscape. Participants in that workshop also noted that sharing tribal knowledge is sensitive. Even the concept of forest management, for example, is anthropocentric and posits that humans are separate from nature—a viewpoint foreign to tribal knowledge-holders. At the workshop, participants noted that the Forest Service still has an attitude of arrogance much of the time, that the agency knows best. One eighty-one-year-old elder spoke of a "simple prescription" that could help make progress between the Forest Service and tribes: open communication, education, respect, and friendliness.

Across the country, tribal members of Miwuk, Karuk, Hoopa, Yurok, Salish, Ojibwe, Washoe, Amah Mutsun, Pueblo, Alabama-Coushatta, and Lakota descent—and others—are all bringing fire back to the landscape. Prescribed fire trainings and exchanges, like the Nature Conservancy's Indigenous Peoples Burning Network, are also helping transmit and amplify this knowledge. Fire brings back more than the historical landscape. It helps rejuvenate a relationship with fire that had stagnated for decades.

"It's easy to understand that when people arrived on this landscape, that it was a landscape that was shaped and maintained by fire and that it was exquisitely beautiful," White says. "It's easy to understand that traditional knowledge of fire should be integrated into fire management practices today, but, man, that's hard to achieve."

In 2021, the Karuk Tribe in California released a landmark assessment called the "Good Fire Report." It opens with sobering numbers:

> In 2020, over four percent of California burned in wildfire. Over 30 people lost their lives in the fires; experts estimate an additional 3,000 premature deaths may have resulted from wildfire smoke. Property damage is expected to top $10 billion. And greenhouse gas emissions from the fires wiped out all of California's efforts to curtail such pollution.

The solution, the Karuk say, is more fire—and fast. Key to making it happen is honoring tribal sovereignty to help tribes restore burning on their land. Fire is a part of life, the report says. Respecting and using it has been handed down as a responsibility, to keep the land healthy for generations to come. It offers a number of policy solutions to help reformulate California's—and the nation's—attitude toward fire. "The time for bold action is now," it concludes.

THE GIFT OF FIRE: BURN PROFILE

On a rare clear August day in northwest Montana, Tony Incashola Sr. and his son, Tony Incashola Jr., walk through widely spaced ponderosa pines in the Jocko Prairie on the east side of the Flathead Reservation. The prairie is just a tiny sliver of the tribe's ancestral land—more than twenty million acres in Montana, Wyoming, Idaho, and British Columbia they've inhabited for at least ten thousand years. Junior and Senior are both members of the Confederated Salish and Kootenai Tribes, and they're visiting this plot of land to check in on a big experiment that showed them how burning could bring back more than just historical forest conditions.

The other important character in this story is something not immediately visible in the prairie: a flower called camas. Sometimes known as wild hyacinth, it's part of the lily family. Its brilliant purple flowers signal a bulb beneath the surface of the earth. That bulb is packed with protein, carbohydrates, and other nutrients. Since they were plentiful, portable, and delicious, they were of particular significance to many Indigenous peoples of the Pacific Northwest. Traditionally, Salish women gathered and baked the bulbs of camas. One woman might have gathered upward of four thousand pounds of the plant a year. It was crucial for the survival of the tribe. The plant thrives in moist meadows—especially in fields newly fertile in the wake of wildfire. Camas, though, has declined since colonization both here on the Flathead Reservation and across the Northwest.

"We were forced to where we're at here today, and we were confined to a certain place, which started to change our way of life and started to change how we gather," Senior says. Traditional plants and food provide much more than nutrition. They're a tether between past and future generations; they're part of tribal identity. In this way, the decline of camas—and the resulting decline of tribal knowledge and tradition—is bound up with the legacy of colonialism. Plants like camas, Senior says, "that's part of the circle of life of who we are." If those plants and knowledge wither away, "then part of that circle is going to disappear."

Absent flame, the prairie where camas used to thrive was overtaken by grasses, weeds, and thick trees. Junior, the head of forestry for the tribes, grew up tromping around these woods—hunting, fishing, gathering, and camping. In those days, the forest looked more like a jungle than the widely spaced prairie of today. It was tangled with dense vegetation, or what Junior refers to as "a ton of stems per acre." As in much of the West, fire in the area had been treated as an enemy for about a

Camas, a flowering plant in the lily family, is known for its bright purple blooms. Historically, it was an important part of nutrition and culture for many Indigenous groups.

century, and the result was a dangerous buildup of fuel for wildfire. What was once open filled in. On top of that, cattle grazed here, further drying out the land, eroding streams, and killing topsoil.

Back then, when this area was an overgrown jungle, there was no sign of camas anywhere. But Senior says tribal elders had always insisted—despite the lack of evidence on the ground—that it had once thrived here. Senior and others in the tribe took note of the damage to the ecosystem. Fire historically burned here every seven to fifteen years. But the area hadn't seen any flames in decades. They reasoned that this area, overgrown and unhealthy, was a perfect candidate for bringing fire back to the ecosystem.

Senior worked for the CSKT culture committee for nearly five decades, keeping traditions and knowledge alive that might otherwise slip through the cracks. A great deal of that knowledge had to do with using fire. When he was young, spending time with his grandma, he remembered seeing smoke in the mountains after a summer thunderstorm. She looked at that smoke and told him that fire wasn't something to be feared; soon, she told him, there would be new growth and new life right where those flames were burning. When there's fire, she said, "the creator is cleaning his room." For Senior, fire was also a regular part of life. On camping trips to this area, Senior's family would often clear a campsite by burning it. That kind of treatment could prepare the land for hunting and harvesting in the future.

Native tribes had an intimate relationship with the flames. They understood its behavior, how it would interact with upcoming rain or snow, and how it could rejuvenate the landscape. Certain tribal members understood these interactions especially well. These people were called "firemakers," or *sxʷpaám* in the Salish language, and—much like a fire management officer or burn boss today—they were the ones responsible for understanding where and when to set fire to the landscape. The fires were meticulous and planned. "They started them to make sure that when it burned, it would control itself," Senior says.

In the Jocko Prairie, tribal members sought to restore fire's cultural and ecological role. It was no easy feat. First, they had to get the cattle out of the area—which involved wrangling owners and animals themselves. Then, they had to deal with the degraded landscape. That meant thinning the area—coming through and cutting away some of the messy undergrowth, especially the smaller trees in between the healthier, older, bigger ponderosa pines. Finally, the real test could begin.

In 2015, on a day when weather conditions were right to keep fire behavior manageable, a CSKT fire crew led by Junior started a controlled burn. They made sure it stayed low and not up in the trees. It was meant to simulate the sort of naturally occurring low-intensity fires that would burn this underbrush at regular intervals and contribute to a healthy forest. The fire left the big, fire-adapted trees standing. The charred bark of the large ponderosa pines did exactly what it had evolved to do—withstand low-intensity fire.

The CSKT burn a few thousand acres every year. About two-thirds of that is near communities, to help slow fire spread and protect homes should a burn break out nearby. But in the Jocko Prairie, the burn wasn't just for protection. It was for rejuvenation. The low-intensity flames burnt away the weeds and nonnative species that had run amok. It cleared the slate for new growth. But it would take months for new growth to emerge from the ashes.

Fast-forward to the summer after the burn. Junior visited the area, and his jaw dropped. "The entire prairie was purple, completely purple," he says. Camas, completely absent from the area for decades, had reemerged from the soil in full force. All it took was one fire. Junior says that with camas back, the knowledge around gathering camas and its myriad uses can flourish too. Cultural events can come back to the prairie. It doesn't stop with camas either. More animals like elk are coming in to use the newly open, lush landscape. "It's not just focusing on one goal," Junior says. "It's all the ecosystem here."

Camas has continued to come back. But much like any healthy behavior, vigilance and consistency are key for the continued benefits of the prescribed burn. For the healthy prairie to thrive, the tribes must maintain the practice every five to ten years. After this experiment, Junior is hoping that the tribes can double the amount of land they burn every year. He says that's tough. But he's optimistic. Part of his role is to bridge the gap between Western science and cultural knowledge. He helps find a sweet spot, where fire can be accepted, not just feared. "Going forward, there is an opportunity someday to really let fire play more of its natural role," he says. "We'll get there."

Junior and Senior describe wildfire—like so many aspects of the natural world—as a "gift." They say it's both a gift to people and a gift to the land. That framing is radically different from the mainstream view of fire. When it's on the news, it's nearly always an enemy—something wreaking havoc that we must put an end to. But imagining fire as a gift rather than a terror helps illuminate the benefits it can bring to a landscape, as well as the people who depend on and interact with it.

Senior says bringing fire to this prairie is one of the tribe's success stories. Today, the serene, widely spaced ponderosas likely look very much how their ancestors would have experienced them. The single purple flower sprouting through the soil symbolizes restoration of ecosystem, tradition, and community. "Fire…is part of who we are and part of nature," Senior says, gesturing at the open landscape around him. "It's part of everything that you see here."

PRESCRIBED BURNING

Indigenous people used controlled burns for hunting, harvest, and ceremony. That was quashed—often violently—by settlers seeking to suppress all flame. Non-Native proponents of so-called light burning thought low-intensity fire could make forests healthier, but their claims were dismissed by the forestry establishment. In the 1960s and 1970s, renegade scientists sought to show the world what "controlled" or "prescribed" burns could mean for the health of trees in some of the nation's most beloved forests. Those ideas took hold for a bit but were by and large stymied by fears of fire. (For more history, see the BURNING section.) But prescribed burning, a millennia-old idea dismissed for more than a century as antithetical to ecological progress, is having a renaissance. Today, scientists, foresters, firefighters, and even community groups say it's a keystone solution to solving our fire problem.

Megafires are ripping through entire forests, causing them to disappear. Communities are ravaged by the flames. One way to think of prescribed burning is like a vaccine that can help the forest stay healthy and survive down the line. Or as preventative work to keep unhealthy forms of wildfire from crippling the ecosystem instead of working to suppress fires that break out and mitigate their effects after the fact. This form of intentional fire, burning with low intensity, emulates what would have occurred naturally on the landscape had it been allowed at its normal intervals over the last century. This kind of fire can remove undergrowth and ground fuel that can spread fire rapidly. It can also lick away at the stuff that could ladder flames up into treetops to create intense "crown" fires in the forest canopy. In a sense, it's a grandiose idea: engineering nature with our own hands and doing it at scale across the country.

Prescribed fires are governed under very strict rules and occur only under incredibly specific, safe weather conditions—and under careful supervision and monitoring. Officials often thin the forest before introducing fire to help keep it in control. If it's too hot, dry, or windy, it's a no-go. A large federal burn can take months or years of planning to get underway. If and when the safety window lines up, the professionals on the ground usually use drip torches, hand-held fire-starting devices that look a bit like gas cans with curly straws. In larger areas or more remote

A drip torch is a tool forestry professionals use to start prescribed fires and pile burns.

locales, they often use ping-pong-ball-looking explosives dropped from the sky, called delayed aerial ignition devices. The flames simmer. They don't jump to the canopy. If the fire "slops over" its containment lines, fire personnel on the scene do everything in their power to contain it.

There's no single, official estimate—but it's fair to say tens of millions of acres need safe, low-to-moderate-intensity fire in the near future. And that's a conservative estimate. Right now, the Forest Service burns only a little over one million acres a year. Even those numbers require some context: between 1998 and 2018, about 70 percent of all prescribed burning took place in the U.S. Southeast, where the legacy of "light burning" (read more about this in the "Suppressed" subsection of BURNING) had normalized the practice. The only federal agency that managed to increase its burning over that same time period was the Bureau of Indian Affairs. Further complicating efforts is this: prescribed burns aren't one-and-dones. The same plot of forest often requires the same treatment over and over, every few years (or more, or less, depending on the ecology of the particular area). In short, right now federal agencies—and especially the Forest Service—aren't keeping up with the need on the ground.

Why is it so hard to get more prescribed fire on the ground? One often-cited reason is public disapproval of the certainty of smoke in the skies. But studies show that the public is actually more accepting of smoke from prescribed fire than many fire managers assume. And yet the country's also never burnt as much land intentionally as it aims to in the near future. In response to smoke wariness, a common refrain in the prescribed fire community is, "How do you want your smoke?" It can come all at once in choking, eye-burning fury. Or it can come slowly, manageably. It will linger in the sky. But it won't limit and threaten health in the ways that keep people in their homes for weeks on end.

Funding, capacity, and bureaucracy are bigger issues. At the federal level, seasonal fire staff is gone in the off-season, when weather generally permits burning. The staff that exists is dedicated to other projects. Fire's no longer top of mind. Permits can be tedious and expensive. Planning can take years, and weather windows might not line up with staff availability. The federal bureaucracy isn't built to be nimble, work quickly, and pivot. Many in the fire community call for more collaboration, more training, and more capacity-building to get more prescribed fire on the ground.

A third barrier is risk. There are few rewards for putting fire on the ground, but lots of punishment if things get out of control. One example came in May 2000. A federal crew ignited a prescribed fire in New Mexico's Bandelier National Monument. The team had hoped to clear grassland that had grown dense with trees due to the lack of fire. It was a dry spring—and later, a review team found that the crew had ignored forecasts for wind gusts. The winds came, and the crew—which was understaffed—couldn't keep up with the flames. The fire escaped and exploded. It ran into the town of Los Alamos, destroying more than two hundred homes and threatening a nuclear weapons facility. The country was astounded at the series of federal missteps in planning and executing the burn. It was "like a rock being dislodged down a hill, leading to a landslide," Interior Secretary Bruce Babbitt said at the time. In one fell swoop, the public and politicians became wary of prescribed burning.

The 2022 prescribed fire season is even more telling. In that year alone, escaped prescribed fires in New Mexico led to the largest fire on record in the state's history. In response to the New Mexico fires, the Forest Service jumped into reactive mode. It put a ninety-day pause on prescribed burning and, after a lengthy review, issued rules that many say will make it harder to get burns done in the future. It isn't the first time that's happened. The public and political outcry made the burning itself the culprit—rather than the overgrown state of the forest. In the fall of that year, another prescribed fire in Oregon jumped slightly out of containment and burnt as much as forty acres of private property; the burn boss was arrested. These incidents gained all the media attention. However,

FIRE STARTS

About 88 percent of fires are human caused. That smaller sliver remaining—12 percent—is caused by lightning. But those statistics are a little deceiving: lightning-caused fires tend to grow larger, burning more land. Lightning-caused fires scorched about 55 percent of the total land burnt between 2016 and 2020.

Also, "human-started" means a whole lot of things. Campfires can get out of control. Dirt bikes and cars traveling off-road can ignite grass and shrubs. Between 1992 and 2015, people started more than seven thousand wildfires on the Fourth of July. And things get a lot weirder than that. One of the worst fires in Colorado's history—2002's Hayman Fire—was started by a Forest Service employee allegedly burning love letters. A nearly ten-thousand-acre fire in California in 2021 was started by a self-described "shaman" who claims to have been boiling bear urine to drink. The four-hundred-thousand-acre Ranch Fire in California began when a man hammered a metal stake into the entrance of a wasps' nest in his backyard, sending sparks flying.

Power lines are a particularly common source of fire starts. Downed power lines in California alone have caused more than fifteen hundred fires over the last seven years. Across the West, power lines have had a role in thousands of fires, sometimes causing some of the deadliest and most destructive burns in decades. Blaming individual behavior doesn't do justice to our fire problem. Industry, development, and our reliance on electricity are also major players in how fires get started.

in aggregate more than 99.8 percent of Forest Service prescribed burns go off without a hitch.

Around the same time, the Forest Service announced an audacious ten-year plan. It hoped to treat fifty million acres—meaning thinning, logging, prescribed burning, or some combination. Twenty million acres it hoped to do on its own land, and it aimed to "support" treatments on the other thirty million acres—which vary in ownership from state to tribal to private.

Dave Atkins, a retired Forest Service ecologist, worked in the agency for more than forty years. "We have this feeling that prescribed burning is really risky. And so people don't want it happening, you know, close to them and they don't like the smoke. But what they don't realize is every time we don't do that prescribed burn, we're just setting it up to burn it in a wildfire when it will be the most severe." Atkins is explaining a classic transfer of risk, from right now, to the future. He suggests that we could learn from other instances where we've been able to change culture and manage risk more thoughtfully: "We decided to put seatbelts in and then later airbags. And, you know, that's what this prescribed burning is the opportunity to do. Develop the seatbelts and the airbags."

Reaching the scale required will take a Herculean effort. But it needs to happen.

THE CHECKERBOARD

Pull up a map of the West on your phone or online. Zoom in. Maybe try western Montana or Wyoming. If you look around enough, you'll start to notice something funky. The land ownership forms a sort of checkerboard pattern. One square is Forest Service—it shows up as "green" on Google—and the next is private. This checkerboard pattern of ownership is a defining characteristic of the West, and it complicates how fires are managed.

The West's checkerboard started with an idea from Thomas Jefferson. In the late 1700s, the way the newly formed country's land was surveyed was, at best, confusing. Inherited from Europe, folks generally used a system of "metes and bounds." Say you're a settler and you wanted to lay claim to your property. You'd walk the perimeter of your land: "Five paces from the creek to the old oak tree, twenty-five paces west to the large boulder…" The boundary lines were far from scientific, uniform, or even permanent. Maps could look downright messy.

Jefferson thought that this idea didn't make an ounce of sense. In the Land Ordinance of 1785, he made permanent what became known as the "Jefferson Grid." This grid system brought a system of standardization—even lines, even-sized plots of land. It divided the landscape into nearly perfect one-square-mile squares. Basically, it started the checkerboard still visible on the map today. The goal was to expedite the settlement process as the country expanded westward.

The next important plot point came in the mid-1800s, while westward expansion was in full swing. Railroads that cut across the nation, many believed, could help the West boom. They could transport goods and people like never before. Starting in 1850, the federal government gave railroads a series of land grants that could help them expand. Those grants generally gave railroads ownership of the land up to six miles on either side of the tracks—but the agreements did so in alternating sections. Railroads would own one square mile of land, and the federal government would own the next. This pattern repeated on an enormous scale. From 1850 to 1871, the federal government ceded a total area of land about sixty times the size of Yellowstone National Park—or about 7 percent of the country—to railway companies. This was all meant to help the railroads fund their enormous construction projects while also encouraging people to move West. The railroads could sell that land to settlers, Congress reasoned. In some cases, that happened. In many others, the railroads sold to timber or mining companies, some of which were subsidiaries of the railroad companies.

In the end, the legacy of railways left us with the checkerboard we see today. And this all links back to wildfire. Those alternating ownership patterns dominate the landscape for enormous parts of the country. But fires don't follow those lines, and that complicates how fires are addressed across huge swaths of land. That could mean, for example, a fuels reduction project stops at a private boundary, or that an ownership line adds complexity to an already tense firefighting operation. A central dilemma of agencies like the Forest Service is how to operate across those checkerboard squares. In essence, they're learning how to erase that checkerboard to promote fire resilience on *all* lands, both public and private.

PRESCRIBED FIRE AND FOREST RESILIENCE: THE SCIENCE

Susan Prichard, a fire ecologist at the University of Washington, understands all the fuss about logging. She grew up on Whidbey Island, a scenic city in the Puget Sound, northwest of Seattle, in the heyday of the industry. She saw clear-cuts as a child and saw how destructive those appeared for the landscape. As a youngster, she fell in love with forests and wanted to protect them. She studied forest ecology in school. As she advanced in her studies—eventually getting a PhD—she discovered that the world of forestry was so much more rich in nuance and complexity than she'd ever expected.

In particular, she gained an interest in how wildfire interacts with forests. She understood the crucial role that wildfire plays in forests. But there was lots of complexity in that: as the climate changes, she learned, it sets the stage for fires to burn so intensely they'd eliminate forests entirely. Warmer, longer summers, for example, can make it hard for young trees to survive. "It can make it or break it for them," she says. She saw dual roles of fire: flames as a force of destruction and change, and as a maintenance tool.

In 2004, Prichard and her wife moved to Washington's Methow Valley. It was an exciting location for them. Living in a fire-prone landscape, she'd get to see the processes she studied firsthand. A couple of years later, the Tripod Fire hit the area. The fire burnt more than 175,000 acres. She says the smoke plume looked like a nuclear bomb went off, and it settled on her community for something like two and a half months. Intellectually, she was excited—"it was like history was in the making right here," she says—but getting that socked in was hard to live with.

When the smoke cleared, she wondered if there was anything she could learn from that huge burn. She'd been studying how forest projects—things like cutting small-diameter trees to thin the forest and reduce fuels, and prescribed burning—influenced the landscape. The theory at the time went that those projects could be a controlled surrogate for the kind of fire that would've naturally burnt through for millennia. She was skeptical of the extent to which these projects made a difference for the forest. And suddenly, she had a natural laboratory of sorts to test out her theory.

Many sections of the burnt forest had been "treated" in the past. Some areas had been "thinned," where smaller-diameter trees were removed to reduce fuel. In other areas, that thinning had been accompanied by prescribed fire. And in others still, there hadn't been any management at all—in scientific terms, control areas. Prichard wasn't sure what those treatments would mean when a fire actually came through: it had often been claimed that these sorts of treatments would help mitigate fire severity. But on the ground, how would the forest fare? Would the fire blaze through anyway? Would any of the widely spaced trees in the treated areas survive?

Prichard says that the results were startling. In areas where there had been both thinning and prescribed burning, about half of trees overall and 75 percent of large-diameter trees—meaning the big, old ones grown large enough to be resilient to the low-intensity burns common to this landscape—survived. "It was like night and day," she says. Areas that had only been thinned fared about the same as areas that hadn't been treated at all. Her analysis showed that the combination of thinning plus prescribed burning could mitigate the severity of a large fire, limit its spread, and provide a reserve of trees in a burnt landscape. The Tripod Fire also showed her that past wildfires directly limited where the new fire could spread.

Fast-forward a few years, and another major fire broke out near Prichard. Called the Carlton Complex, it burnt as much land in just a matter of days as the months-long Tripod Fire did. In all, it torched more than 250,000 acres, and it was a fire of almost unimaginably extreme proportions, the kind of devastating, uncontrollable fire the West is seeing more and more of. Since the fire was so intense, she thought she might get different results. Perhaps the fire was so powerful, it would scorch right through even the areas that had seen prescribed burning and render all that work and management meaningless. After trumpeting the efficacy of prescribed burns after her study of the Tripod Fire, she says, "I thought I was going to eat some humble pie."

Once again, she looked at areas that had been thinned, areas that had been thinned and seen prescribed fire, and spots that had been left untouched. She ran the data. When she got her results, she was surprised. Once again, thinning and prescribed burning, combined, helped quell the severity of the fire. Her results held whether her team analyzed total tree survival or that of just older, large-diameter trees. While the results weren't quite as stark as in her study of the Tripod Fire, she also learned that those forest projects could be particularly effective on leeward slopes, or in the shadow of the wind.

Prichard is not the only person who's studied this. A 2022 study looked at the Dixie Fire, which scorched nearly a million acres in California in 2021. Studying other fires, the authors hypothesized that fire-prone landscapes show a sort of "ecological memory." When fire burns, it follows the severity of where fire burnt in the past. So an area that experiences low-intensity fire is likely to experience it again. In the Dixie Fire, the authors saw a test case: Would enormous wildfires burning in extreme weather still demonstrate that same sort of "memory"? Just like in Prichard's study, the authors identified past fuel treatments in the Dixie footprint. They also controlled for changes in weather and mapped past burns. Even a fire as enormous as this one, they found, burned with a "strong ecological memory" of past blazes. Areas that had burned at high severity in the past burned at high severity once again. Fuel treatments like thinning showed little to no difference, just like in Prichard's studies. They concluded, "The Dixie Fire is a particularly dramatic example of the significant potential of low severity fire to blunt undesirable fire effects in successive fires in an era of increasing large and severe wildfires burning under extreme conditions."

Now, Prichard says, "the science is really clear." With coauthors, she reviewed thousands of peer-reviewed papers. There's nuance, but the basics are somewhat similar: prescribed burning in thinned forests makes a dent in wildfire severity. That point is important. Thinning alone—which is often much easier to get done—is generally insufficient. In many studies, there's no statistical difference in thinned forests compared with just leaving the forest alone. It's also crucial to note that the kind of forest treatment that's most beneficial to the landscape, and where, varies tremendously from one forest to the next. But the upshot is: one of the best ways to fight the most damaging effects of fire is with more fire.

These results also don't suggest that thinning is always insufficient. In addition to fire resilience, thinning can bring other benefits. A thinned forest can provide more open habitat for wildlife. It can allow more sunlight and nutrients for the larger remaining trees to survive. It all depends on context. Thinning projects aren't designed to stop big wildfires. They're designed to help reduce the fuel load and can also help firefighters find a foothold in otherwise dense tree stands.

Prichard sees logging, thinning, strategic logging, and, especially, prescribed burning as tools in the toolkit. The challenge, though, is reaching the scale we need. "We've really gotten ourselves in a hole right now," she says. For now, the Forest Service's goals will be hard to meet. But more thinning and burning, in strategic locations, will be crucial in helping both communities and forests become more resilient to extreme fires. Getting that necessary work done requires stakeholders who have butted heads for decades to find common ground. It means finding compromise, and short-term risk for the long-term good of communities and ecosystems.

So, how do we find a collective vision for how we want our forests to look and feel? And once we figure that out, the crux is: How do we get it all done? Understanding some of the hurdles ahead requires getting a sense of the evolution of forest projects like prescribed fire and thinning within the Forest Service.

FOREST PROJECTS:
LOGGING VERSUS ENVIRONMENTALISTS

Work that can increase resilience to wildfire across the West is bound up with passions around resource extraction and how forests ought to be used.

That's in part because in the 1980s, a war began over trees in the Pacific Northwest.

On one side were environmentalists, who saw destruction in forests all around them and wanted to do something about it. On the other were loggers and timber companies, who supported entire communities with their work in the woods.

Tensions over the fate of the country's forests had been simmering for years. At the heart of the controversy wasn't wildfire; it was logging. Wood production in national forests was at an all-time high. Timber harvest peaked in the late 1980s at over twelve billion board feet per year. One board foot equals a piece of wood one foot long, one foot wide, and one inch thick—so if you lined up all the wood harvested from national forests in a single peak year end-to-end, it would circle in the earth along the equator almost one hundred times. The Forest Service, after all, was created under the Department of Agriculture to sustainably grow and harvest trees. Timber sales generate revenue for the agency.

REFORMULATING FIRE SEASONS

Just about every year, folks talk about the number of acres burnt by fires. (We've done it many times in this book.) More than ten million acres were scorched in 2020 and 2017; more than nine million were left in ashes in 2006, 2007, and 2012. Whenever the acreage breaks a record, headlines pop up across the nation. But acres burned, or not burned, can obscure a lot of important stuff: How severe was the fire? How many homes were lost? What did it do to the watershed and air quality? What ecological benefits occurred? How did it affect future fire risk? A million-acre fire could be a really good thing, provided it occurs in the wilderness with minimal impact on what the Forest Service calls "values at risk." On the other hand, a one-acre fire can be devastating if your home sits on that acre.

We need to shift how we quantify fire. We need more fire, of the right kind, in the right place. And to make that happen, we need new metrics. One option is to explicitly focus on how many people are evacuated, the severity of fires, and the number of structures lost. That could include quantifying things like prescribed burns—and also wildfires that simmer in large areas without (or with relatively little) human intervention. In short, we need to set goals on the outcomes we want for both ecosystems and communities. Then, we can start to track how we're actually getting there.

In 1970, Arnold Bolle, dean of the University of Montana's forestry school, along with several colleagues, looked at how forest projects had been conducted in the Bitterroot National Forest, south of Missoula. The study, known as "The Bolle Report," found that even though the national forests were supposed to be managed for "multiple uses"—including recreation and wildlife—producing timber was above all the overriding bottom line of the agency's activities. It found that the agency over-relied on clear-cutting and, crucially, that it cut the public out of most of the operations.

The report, along with another controversy over clear-cutting, this time in the Monongahela National Forest in West Virginia, paved the way to begin a change in the agency's attitude toward forest management and clear-cutting. It also heightened public mistrust toward the agency's motives. That mistrust echoed across the country, as the public felt ignored and abused by the Forest Service.

Early on, activists began "tree-spiking," hammering nails or other bits of hard metal into trees slated for logging. When a chainsaw would bite into the flesh of a tree, it would hit the nails, throwing the saw's chain off its bar and making it potentially unusable. As the chain went flying, it could cause injuries to the person holding it. At least one timber employee was nearly decapitated when his saw hit a spike in a tree he was cutting in Northern California in 1987.

Looking for a less injurious way to prevent logging, environmentalists got a different radical idea. In the mid-1980s, logging was proposed in old-growth Douglas fir in Oregon's Willamette National Forest. One way to block it, they reasoned, was with their bodies. Loggers wouldn't cut down a tree with somebody in it. So began an era of "tree sitting," or climbing up to literally live in treetops, sometimes for months on end. The most famous tree sitter lived up in the branches for more than two years straight.

Activism like tree sitting and spiking continued. So did rallies and protests. Protestors blockaded roads, chained themselves to trees, and buried themselves in rocks in the way of machinery. They were routinely arrested, and tensions were escalating. Just before a months-long parade of protests known as the "Redwood Summer" of 1990, a car bomb went off under the seat of two prominent activists. The bombing remains unsolved—but authorities at the time sought to prosecute the two victims of the explosion, who survived. Another partially detonated and similar bomb was found at a sawmill. The real culprit remains a mystery.

While most protests were peaceful, there were also reports of protestors beaten and harassed by loggers, or having eggs or rocks thrown at them. Loggers, too, reported being terrified of showing up to work each day for fear of what could happen.

The sides of the battle represented two competing value systems: one sought to preserve trees they felt could never be replaced. The other viewed trees as renewable resources. They grow back, after all. And we need wood to build homes and continue our way of life. Neither side was entirely right or wrong. But at least at the time, they couldn't find common ground.

A SMALL BIRD AND THE DECLINE OF LOGGING

As logging continued relatively unabated, members of groups opposed to the scale at which forests were disappearing across the Northwest formed a theory. Spearheaded by the National Wildlife Federation, the thought went that maybe saving a forest could start with saving an animal that depends on it.

That charismatic mini-fauna came in the form of an owl. The northern spotted owl has a wingspan of about forty inches. Like other owls, it perches in the limbs of trees and preys on mice and other small rodents, mostly at night. But unlike other owls, it depends on old-growth forests to survive. By the 1980s, as logging was thriving, northern spotted owl numbers in the U.S. were dwindling.

Today, massive, forest-destroying megafires in the Pacific Northwest threaten spotted owl habitat. But back then, the country by and large behaved like it had wildfire under control. In 1988, the enormous, headline-grabbing Yellowstone fires flared up. The conflagration burnt more than a million acres in and around the country's first national park. Much of the public—especially those on the coasts, far away from the action—saw only destruction and devastation. Public policy toward fire was in many ways pushed toward further suppression, even though there was burgeoning science showing fire's beneficial effects for the landscape.

The other major player here was a piece of legislation passed with overwhelming bipartisan support under President Richard Nixon: the Endangered Species Act. The ESA was meant to help bring back the nation's once abundant wildlife. Today, it's credited with helping critters like the bald eagle, peregrine falcon, even the American alligator come back from near extinction. When a species is listed as "threatened" or "endangered," though, that classification includes a designation of "critical habitat" that the species needs to survive. When an animal receives federal ESA protection, it means big swaths of land can be off-limits to development and industry. Environmentalists opposed to logging soon started a long and strategic battle to get the northern spotted owl protected by the ESA.

Finally, after years of studies, debate, and lawsuits, it happened. In 1990, the northern spotted owl was listed as "threatened" under the Endangered Species Act. This was a critical plot point in the saga of logging and environmental protection  in the U.S. The timber industry was worried that protecting the spotted owl would mean next to no logging could happen in the Pacific Northwest; the creature depended on old-growth forest to survive. That could mean tens of thousands of jobs—an entire industry—would be left in the dust. Many timber workers felt that environmentalists cared more about an animal than the well-being of people. Bumper stickers, for example, read KILL A SPOTTED OWL—SAVE A LOGGER, or SPOTTED OWLS TASTE LIKE CHICKEN.

Over the years that followed, as additional lawsuits played out, the Forest Service set aside millions of acres that would be protected as spotted owl habitat. The timber industry as a whole had already seen big declines since World War II. In the years since the spotted owl gained federal protections, the industry saw drastic reductions across the Northwest. At the industry's peak, trees harvested on Forest Service land made up about 11 percent of overall timber produced in the country. Today, that number is down to less than 3 percent. Annual harvest hovers at a little less than three billion board feet per year—less than a quarter taken during the industry's heyday three decades ago. Entire towns reliant on timber continue to struggle to find their identity after the industry's decline. The decline, of course, wasn't only dependent on lawsuits and the ESA. Other factors influencing the market and jobs, including global recessions, price fluctuations, and mechanization, also affected the amount of timber harvested on federal land.

Around that same time, the South Canyon Fire blew up in Colorado. Fourteen firefighters died, and the federal government updated its wildfire policy. That's something it hadn't done in decades. Where possible, the report said, "wildfire will be used to protect, maintain, and enhance resources." As the primacy of logging was in decline, a spotlight was shifting onto the broader ecological system at play in forests, and wildfire's role in it.

The story of the spotted owl informed a change in forest politics—and how the country's forests are made more resilient to wildfires—that lingers to this day.

FOREST MANAGEMENT TODAY

"Radical environmental groups would rather burn the entire forest than cut a tree or thin the forest," Ryan Zinke, former secretary of the interior, told reporters in 2018 as wildfires raged in California. The statement made headlines across the country—and it's a narrative that's taken hold: across the West, politicians and folks in the forest industry routinely blame litigation, often brought by a relatively small group of passionate conservationists, for holding up proposed forest projects like thinning trees, clearing brush, or prescribed burns. We can manage the forest in careful, systematic, and science-based ways. And that often involves removing trees. Or, the argument goes, we can let nature do that for us, as wildfires grow uncontrollable and whip through forests across the West. Today those spotted owl bumper stickers have been updated. Across the West, you might find bumper stickers that read: Log it or watch it burn.

Instead of logging and clear-cuts, the language that gets used now involves terms like *fuels reduction*, *improvement cuts*, and *thinning*. Some environmentalists say that it's doublespeak, that it's the same old attitude of trying to cut down more trees, but now under the guise of wildfire resilience. They point to projects that "thin" small-diameter trees too small for commercial value in some areas, but clear-cut farther into the forest where the trees are more profitable. In some of these projects there might be a prescribed burn—something important for fire resilience—but coupled with a logging project elsewhere in the forest. Or projects treat stands of timber deep in the woods, far from the "values at risk" they claim to protect. The Forest Service gains revenue from logging contracts, many allege, and so has an incentive to continue to focus on the bottom line, meaning timber harvest, especially with tight federal budgets. It's important to point out that, while a great deal of science shows that thinning forests alone doesn't offer much protection when extreme burns come through, thinning projects aren't only meant for fire resilience. Those treatments can offer potential safe havens for firefighters during a burn and can help remaining older trees thrive.

The weapons that would change the tides of the war on logging weren't tree spiking or sitting high in tree branches or blockading roads; what proved more effective were lawsuits and bureaucracy. In addition to the ESA, the other major character to understand in how logging and forest projects happen today is the National Environmental Policy Act. NEPA is meant as a review process that makes sure federal projects don't bear significant environmental costs. That guarantees that logging projects, depending on their size, generally require Environmental Impact Statements or Environmental Assessments, lengthy documents—sometimes years in the making—that analyze the impact of projects on ecosystems and wildlife. They also give multiple windows for public input. Those are times when you, as a community member and member of the public who technically owns part of national forests, can make your voice heard. NEPA plays a key role in legislation related to forest projects. And, like any bureaucratic process, litigation can slow it down.

In one sense, it's easy to see logic on both sides: as timber harvests have drastically gone down in national forests in the West, wildfires have skyrocketed in terms of size and severity. In 1990, the Forest Service spent more money on logging than on anything else. Now, the agency's budget is dominated overwhelmingly by wildfire. About a third of lumber in the U.S. goes to building new homes. Many of those new homes are in the WUI. Lawsuits over forest projects continue to this day. Some conservationists have taken the strategy tested by the lawsuits over the listing of the spotted owl and run with them: they allege projects build too many roads, cut too many trees in old growth, and ignore important habitat for the spotted owl and other federally protected critters, like grizzly bears, Canada lynx, and bull trout.

The tension here has created a bit of a stalemate. The Forest Service seems stuck between two worlds: that of resource extraction and of building science-backed resilience in the face of a changing climate. And complicating all of that is the tough project of communicating forest decisions to a divided public. The federal government wants to increase the pace and scale at which it conducts forest projects. But it also cites "analysis paralysis" and litigation as a factor holding back getting the work done. One review of NEPA and fuels reduction projects found that it takes an average of nearly four years from the time a thinning project is initiated to when work begins on the ground. For prescribed burning, that timeline is even longer: nearly five years. Projects bigger in scale that require even more environmental review (in technical terms, an Environmental Impact Statement) tack on an additional two to three years to that timeline. Public input on projects like these is a key part of our democracy and environmental policymaking and takes time.

It's one way in which the uniquely modern form of political and cultural polarization is playing out. It also reflects the Forest Service's legacy of distrust in many communities. A review of all lawsuits brought against the Forest Service from 1989 to 2008 found that plaintiffs won or settled 51 percent of the time. In more recent years, those numbers look slightly different: between 2005 and 2018, the Forest Service won about 67 percent of cases. But even though the Forest Service is winning more cases, many people in the conservation world say these statistics point out an important point: much of the time, the agency isn't playing by the rules, and lawsuits are essential to accountability.

The fear of lawsuits and blame toward environmentalists might be overplayed. The data show that across the country from 2005 to 2018, only about 2 percent of Forest Service projects requiring Environmental Assessments under NEPA were litigated. That number rose to 17 percent for projects requiring Environmental Impact Statements, which tend to be much larger in scale. Those numbers are much higher when you zoom in on the West and forest products and fuels reduction projects in particular.

Wherever you fall along the fault lines here, the battles continue. Lawsuits, public statements, and op-eds continue for forest projects in national parks, national forests, and other public lands. In some cases, tribes are demanding a say in forest management, or for "land back." Protestors, sometimes going by the term *forest defenders*, continue to organize demonstrations and blockade roads in the U.S. and Canada on certain land sales. Adding complexity today is that the stuff that needs to get removed from forests isn't the thick, valuable trees that timber companies and sawmills may salivate over. It's the smaller-diameter, younger growth that's sprouted up in the wake of fire suppression. That vegetation has little—if any—commercial value. (The Forest Service and other organizations are working to find new technology and ways to generate revenue from that timber.)

These projects by and large take place on public land, and you are partly an owner. You can help inform how and where these projects take place in the future.

FOREST COLLABORATIVES

In the mid-1990s, a group of environmentalists wearing ski masks filed into an Ashland, Oregon, ranger station. The Forest Service had proposed removing some trees, partly to reduce fire risk. A shouting match ensued, and the activists delivered a note, saying that the agency shouldn't cut down any trees: "Diplomatic channels have been exhausted. Consider yourself warned..." At a particularly rage-filled forest meeting in Montana's Swan Valley, an attendee threw a tobacco tin at a speaker's head. The sides of the debates over forest management were clear. Nobody was giving any ground to their purported enemies. But the modern era of megafires has changed how both sides see risk in our forests, and how they relate to one another.

Enter "forest collaboratives." These are forestry committees composed of diverse groups from a variety of perspectives. They might include folks from a small environmental organization and a big group like the Nature Conservancy. They sit alongside people who live in or near the forest in question, members of the Forest Service and state forestry agencies, ranchers, and representatives of logging companies. Just one or two decades ago, meetings like these might've been a recipe for a black eye, but today they're offering a new approach to forestry.

In Ashland, where those masked environmentalists threatened the Forest Service, the Ashland Forest Resiliency Project has taken the helm of forest projects in the area. The project managed to bring together the same folks who were butting heads back in the 1990s to reduce the area's vulnerability to fire. Same goes for Montana's Swan Valley. There, the Southwestern Crown Collaborative has managed to treat about twenty-eight thousand acres in the WUI, and another eighteen thousand acres outside it.

Today, forestry collaboratives exist in states all over the country—including, in the West, at least Oregon, Montana, Washington, California, Idaho, Colorado, and New Mexico. These

UNDERSTANDING FOREST PROJECTS

Logging, thinning, and burning projects aren't going away, and the Forest Service is hoping to ramp up the scale and pace of its work to help landscapes across the West become more fire resilient. That requires balancing the social and aesthetic value of forests with the science of how and where forest projects can be most effective at wildfire resilience, and getting the labor necessary to actually do the work in forests.

How it all gets done depends in large part on how the public and communities nearby perceive and advocate for the places that they love and live in. Public participation is key to the process as more forest projects take shape. The terms below can help you stay informed and give your input when these projects come to your neck of the woods.

Fuels reduction any process in which things that can increase fire hazard are removed from a forest. Often, this comes in the form of thinning, or removing small-diameter trees that would've burnt in historical, low-intensity fires.

Regeneration harvest a category of harvesting trees that can take many forms. This often means a clear-cut, but can mean any sort of forest project that sets the stage for either natural regeneration, by growing naturally dispersed seeds, or artificial regeneration, through planting and seeding.

Seedtree cut a harvest in which all but a few trees are cut. The remaining trees are meant to help regenerate the forest in the landscape later.

Shelterwood cut a logging method that removes up to half the mature trees from a stand, leaving behind enough mature trees to provide shelter and seed for new trees to grow. Once the new trees take hold, the remaining mature trees are removed. This shelterwood method allows for healthy and progressive regeneration within a forest.

Improvement cut a forest harvest in which individual trees are removed based on certain characteristics. They could be diseased or damaged, or this practice could help build more diversity of age class and tree thickness into a given stand.

Old growth a stand of forest characterized by large, old, often never-cut trees. In addition to ecological benefits, many proponents of leaving old-growth forests out of logging projects claim that they're key to sequestering carbon and helping the fight against a changing climate.

Mechanical thinning using heavy equipment to reduce the density of trees in a stand. This often involves large, cranelike vehicles, piling brush and slash, and creating fuel breaks. This helps make a forest more open and is one step to mimicking historical wildfires. It also allows firefighters easier entrance should a wildfire break out nearby.

Clear-cut a method of logging in which all or most trees in a stand are uniformly cut down. Clear-cuts are commercially attractive, but the effects on water, soil, and habitat can be devastating.

Plus, for many users of the forest, they're an eyesore. Clear-cuts have ramped down in frequency over the last couple of decades, but many land managers argue small clear-cuts can help restore forest conditions in overcrowded and unhealthy plots of land.

Commercial thin selectively harvesting specific trees in a forest for sale as wood products. Thinning can reduce the competition for water and nutrients in a forest and increase the health of remaining trees. It also can make the forest more resilient to wildfire by reducing fuel and creating breaks in the canopy.

Precommercial thin selectively cutting down the unhealthy, young trees in a forest. Typically, these trees are twenty years old or younger. Removing these trees allows the remaining trees to thrive and grow, as they will have greater access to water and nutrients and are more likely to resist disease and insects.

Small-diameter trees trees less than twelve inches in diameter. These smaller trees are typically uneconomic and inefficient for commercial harvest, but they can clog a forest and increase fire risk. Uses for small-diameter trees include engineered wood products, mulch, pulp chips, and particleboard.

Critical habitat areas essential to the recovery of species protected by the Endangered Species Act. Federal agencies doing projects on critical habitat must consult with the U.S. Fish and Wildlife Service to ensure that work doesn't adversely affect species.

Endangered Species Act enacted in 1973, the ESA mandates protections for fish, wildlife, and plant species that are designated as threatened or endangered. The act specifies a process for establishing, managing, and removing these designations.

National Environmental Policy Act a landmark law enacted in 1970 that established a national environmental policy. NEPA requires all federal agencies to evaluate the environmental impacts of their actions.

Environmental assessment an analysis process established and required by NEPA, designed to assess whether a proposed federal action will have a significant impact on the environment. If an EA determines that significant impact is likely, an environmental impact statement is required.

Environmental impact statement a formal, structured process that documents the positive and negative environmental effects of a proposed federal action. An EIS is required for any proposed action for which significant environmental effects are expected. While EAs are for projects that are smaller in scope, an EIS is for large projects that might affect entire landscapes or ecosystems.

Categorical exclusion a type of action that's been deemed not to have a significant impact on the environment. Categorical exclusions get fast-tracked through the NEPA process and require far less review.

Ecological succession the process through which the mix of species in a particular ecosystem changes. Succession typically occurs after a disturbance like a wildfire or flood.

Age classes groups of trees in twenty-year age increments. Even-aged stands are made up of trees from a single age class, while uneven-aged stands consist of trees from three or more age classes. Uneven-aged stands are more diverse and are considered in balance when the age classes within them are equally represented.

Scoping a specified period during the beginning of the NEPA process in which members of the public can tell the Forest Service or other federal agencies about primary concerns they would like to see addressed in an EIS.

Public comment every EIS or EA under NEPA has multiple time periods to offer input. That starts with public scoping but doesn't end there. Public comment is also allowed, usually in a more limited context, after draft and final versions of the documents are published. Public comment can be written and submitted to the federal government, but it can come in other forms, too, including in workshops, public meetings, conference calls, or hearings. Drafters of NEPA documents read and respond to comments submitted by the public.

LOGGING AFTER THE FIRE

After a wildfire torches an area, the Forest Service and other agencies often want to minimize their economic losses. All those burnt trees, after all, are worth some money. So state and federal agencies *salvage log*, or remove standing dead or near-dead trees.

From the agencies' perspective, salvage logging can be critical to mitigate the severity of future fires by removing fuel, and getting rid of the near-dead trees can help make the forest less vulnerable to potentially devastating insect outbreaks. Some recent laws have allowed fast-tracking projects like these.

Critics say some of the trees that are taken in "salvage" projects will recover naturally, and that this is one more way for the Forest Service to meet its board feet quota. Those burnt trees can help prevent erosion after a fire, by holding dirt in place that might otherwise be swept into streams. They also provide critical wildlife habitat, especially for critters like cavity-nesting birds.

The science doesn't show total consensus, but one analysis of about one hundred salvage logging studies concludes that the practice "does not necessarily prevent subsequent disturbances, and sometimes it may increase disturbance likelihood and magnitude." Another study was more blunt: "The word salvage implies that something is being saved or recovered, whereas from an ecological perspective this is rarely the case."

Where a salvage logging project is proposed near you, attend public meetings and talk with your local foresters to find out where, how, and why salvage logging projects are in the works.

collaboratives have mostly emerged in the last couple of decades. While forest projects had divided stakeholders for decades, wildfire served as a force that could, in many ways, unite. No one wanted enormous, stand-replacing fires that threatened homes and property. Everyone wanted resilient forests. They just had differing ideas on what that means. Congress started funding collaborative conservation in 2009, and since then that federal program—which funds only certain large-scale collaborative projects—has treated forest projects all over the country. The scale that's been reached is huge. If you bundle all the work up, it equates to a chunk of land the size of Pennsylvania and Rhode Island combined.

Not all collaboratives are successful, and critics say that they're collaborative in name only. They allege the groups prioritize the "Big Greens"—or the environmental groups with the most money and resources (and most willingness to "sell out"). They also claim that the groups homogenize the perspectives at the tables and often overrepresent groups with an interest in resource extraction. They also allege that they elevate the voices of folks who live nearby, who might benefit economically from logging. Public land is owned by everyone, those critics say, not just people close to the forest. Critics say broader representation is needed to give adequate voice to the land and wildlife.

These groups often create frustration and exasperation, but also genuine exchanges of ideas. Confronted face-to-face with folks who were once enemies, truly novel ideas can emerge. People understand one another on a human, not just ideological, level. It's a path toward empathy, even if not necessarily consensus. Two University of Montana scholars, Martin Nie and Alex Metcalf, write that "instead of completely abandoning collaboration in principle, our hope is that those critical of these processes join them in order to fix what they perceive as being wrong with them." They also contend that collaboration ought not be an alternative to litigation or the NEPA process. Rather, it's a starting point for meaningful public engagement in forest projects.

In a world characterized by division, forest collaboratives are one way to bridge gaps and find common ground. In terms of getting them started, members of one particularly successful Oregon-based collaborative told *New York Times* columnist Nick Kristof semifacetiously, "It helps to have alcohol, and it helps to have food."

IS THERE ANOTHER WAY?

Where Lenya Quinn-Davidson, a fire adviser with the University of California Cooperative Extension, grew up in rural northern California, fire was all around her. Burning trash and debris in the backyard was the norm. In the public-land-dominated landscape, wildfires were common; her mother spent summers working as a caterer at fire camps. The mission, when it came to wildfire, was clear: put it out. "The narrative around fire was very adversarial," she says. Davidson's career would go on to challenge that narrative and instigate a movement that has the potential to shift where prescribed burning occurs, who does it, and at what scale.

When Davidson was in college, her relationship with fire began to shift. She learned about its role in ecosystems. She became fascinated with the potential of fire to heal. This was the mid-2000s. Prescribed burning was barely mentioned in the mainstream fire community. Still, Davidson tried to promote the concept and its implementation. She held Prescribed Fire Training Exchanges, or TREX, to build capacity. She was part of the Northern California Prescribed Fire Council, which she says was the first council in the West to convene to discuss the need for more prescribed fire and the barriers to getting that work done.

As a university extension agent, it's Davidson's job to interact with the public. She got question after question from landowners nearby saying: "I want to do a prescribed burn on my property, but I don't know how." At first, she didn't have any answers for them. Cal Fire, the state agency in charge of fire and forestry, was a "dead end street," she says. The department just didn't have the capacity or will to get the work done. But then she came across something called Prescribed Burn Associations, or PBAs—in the Great Plains of all places. These are community groups who get together to burn land. The idea, Davidson reasoned, was kind of radical—and it had potential to reshape how burning gets done in California. So in 2017, she and her colleagues went to Nebraska, to learn from groups of landowners having success bringing fire to their own properties. "It was like, oh, of course, community members getting together and helping each other out," she says. "It's like a barn raising. It's like a branding. It's things that we do in society already, but it's with fire."

She took that idea home with her and started a PBA locally, in Humboldt County, the first in California. It was staffed entirely with volunteers, and in the first year they were able to get some big projects done. The momentum began to grow and grow, Davidson says. With her colleagues, she spent two years traveling the state, explaining the permitting process, how to plan, how to follow the law—the nuts and bolts of how to get the work done.

As her idea for PBAs in the West was taking shape, the 2017 fire season hit. Fires in California's wine country north of San Francisco killed more than forty people. "No one had ever seen something like that before," Davidson says. The next year, the unthinkable happened again. The Carr and Camp Fires were responsible for the deaths of more than ninety people. Nearly twenty thousand buildings were destroyed. Those catastrophes spurred Californians to think differently. Suddenly, prescribed fire became part of the state and national conversation about solutions.

But there was still a problem: the government was acting too slowly. Davidson didn't want to wait around for the government to change; there was no time for that. She figured it was time to shift the focus away from the feds. After all, she says, the West has been dominated by what she calls a "fire suppression culture" for a century, yet the assumption is that the only people who are qualified to manage fires are qualified to put them out. That forms a central paradox of fire management in the West: "We're asking the people who are putting fires out to also be the ones who are restoring fire," she says. "And they can't do it." Davidson wanted to get ordinary people involved in getting fire on the ground. Maybe, she thought, they—the people who actually live with the risk—are the perfect folks to be restoring the ecosystem.

The vision here can sound idyllic, but the reality of spreading PBAs even wider has some major barriers. First and foremost is the law. What happens when a fire *does* escape comes down to liability. It's useful to break down legal liability into a few categories. States that conduct the most prescribed burns have "gross negligence" standards for both suppression costs of escaped prescribed fire and third-party damages. That means that whoever put the fire on the ground would have to be proved "grossly" negligent; they would've had to do something seriously wrong to have to pay for the damage. A slightly more limiting standard is "simple negligence" laws. That means burners have to prove they took reasonable care in lighting the fire. The harshest laws on the books are "strict liability." That means if anything goes wrong, whoever set the fire is on the hook. Strict liability laws can be a major barrier to prescribed burning. However, many states haven't set standards for prescribed fire liability at all, so there's a huge question mark over what would happen should anything go wrong. Anyone interested in setting prescribed fire certainly wouldn't want to find themselves serving as the test case in the courts that could inspire further clarity.

It's worth pointing out that the escape rate of prescribed fire is less than 1 percent. Escaped prescribed fires that cause damage are an even smaller percentile. One 2020 study looked at twenty-three thousand state, federal, and private prescribed burns. Across all those intentional fires, only one insurance claim was ever filed. But getting the right laws in place can help burners feel more protected in their efforts.

Another issue is insurance. Folks who intentionally set fire need insurance should something go wrong. In California, for example, insurers are backing out of just about all aspects of the fire market. But the California legislature set aside $20 million in a prescribed fire claims fund. Should a prescribed burn get out of control, the burners could apply to use those funds to reimburse damages.

Legal barriers aside, Davidson wanted to turn the traditional model of prescribed fire on its head. She devoted her efforts to helping ordinary citizens get involved with PBAs. As the need for more fire on the landscape went mainstream, Davidson's effort took off. In four years, the Humboldt County PBA has managed to burn more than twenty-seven hundred acres. By 2022, California had more than twenty different PBAs, all working to burn around their own communities.

"These practitioners, often just in blue jeans and ball caps and burning with their children, grandparents, and neighbors, are reimagining prescribed fire in California," Davidson wrote in an essay. The idea is both simple and radical. She calls it a social movement: community members lighting fires in the areas they care about the most. It's a shift away from the emphasis on federal decision-making and power that so often dominates the discussion about wildfire. It also cuts through a great deal of the red tape that bogs down federal projects. The work is mostly on private land, done by volunteers, so it simplifies permitting and environmental review. Participants don't have to get state or federal funding in many cases. Plus, the work that gets done is the work that's most meaningful to community members themselves: the land immediately surrounding a home. A forest directly abutting a community. And Davidson's PBA program is replicable—she's helped form a network of practitioners, where each group can learn from the experience of others. These groups are also nimbler than government agencies. "Anytime there's a burn window, as long as you have someone to coordinate, you can get people out to burn," she says. "There's always someone available." Where the government has struggled, community members can thrive. And the concept is spreading beyond just California; new PBAs have sprung into existence in Oregon and Washington too.

Davidson's vision is so much more than just PBAs. As home-owners or renters living in fire-prone landscapes, it's easy to expect others to protect us from wildfire, whether those others come in the form of the Forest Service, state agencies, or even the local fire department. PBAs offer a version of restoring ecological health and mitigating risk that necessitates a more engaged public, and a public with agency in building collective resilience to fire. This is part of a bigger idea of hers. She says the idea that people are separate from nature—that we hold dominion over the natural world—got us to where we are today. Davidson says that's a sort of colonial mentality, and it's present both in the culture of fire suppression and among environmental activists who don't want to see trees cut down. People have always been a part of the natural world. We've always molded it, at least to an extent. Finding a sustainable future with fire requires letting go of the idea that we can separate ourselves from the natural processes around us. "I just really love the idea of breaking down the walls between people and nature," Davidson says. "We have a role. And it can be positive."

If you want to start or get involved in a PBA in your own community, reach out to other practitioners. Find a network of California PBAs and learn more about the process at https://calpba.org/.

WHAT YOU CAN DO

- Write your local state and federal representatives. Tell them what you think needs to happen in the forests near you.

- Get in touch with the nearest ranger district or section of national forest to learn more about forest projects in your neck of the woods.

- Find out about forest collaboratives near you. Get involved.

- Right now, many federal and state resources are focused on forest projects like logging, thinning, and prescribed burning. Many think we should allocate those resources toward more support for retrofitting homes for fire resilience and projects near communities instead—or, making sure both approaches are part of the toolkit. What do you think? Reach out to your local representatives and Forest Service and let them know where you stand.

- If you live in an area with a Prescribed Burning Association, reach out to see how you can help. If there are none nearby but you think there should be, reach out to other practitioners to learn how you might go about helping organize in your area.

- Get your voice heard. Participate in the public comment process on proposed forest projects. Attend meetings. Learn as much as you can. Tell your friends and neighbors to get involved too.

- Focus on your own home and community. See what work you can get done on the areas you care about the most. (For more on that work, see the ADAPTING section.)

ADAPTING

It's easy to feel like wildfire is an out-of-control, unsolvable problem. But there are ways to keep yourself and your neighbors safe. These solutions are within our grasp, with resources and technology that exist now. Frankly, many of these solutions are actually quite simple. But finding a fire-resilient future isn't easy. It requires action on the individual, community, and national levels. It requires addressing climate change at the same time that we reframe our relationship with flames and taking responsibility for where and how we've built, and it requires identifying exactly where the wildfire problem lies so that we can take the proper action.

HOME

When homes and communities burn, the images we see in the media are dramatic. Stands of trees going off like matchsticks, sending massive flames into the sky. Wildfire, those images convey, is a deadly force that sweeps down from the hills and overtakes homes. That narrative, though, leaves people with little to no agency over the situation. Communities are helpless victims of the overwhelming force of fire.

Decades ago, this is the image that Jack Cohen assumed was right. Nowadays, Cohen is retired, but his work as a researcher for the Forest Service revolutionized how we think about homes burning in wildfires. Cohen grew up around fire and always paid close attention to the flames. Early in his career, he didn't think much about homes and structures. He assumed that when they *did* burn, they were simply overtaken by the fiery front of flames, like an enormous wave crashing on a beach.

One particularly destructive fire on which he cut his teeth was the 1980 Panorama Fire. It started as arson, but ninety-mile-per-hour winds swept it through more than three hundred homes, killing four people. At the time, wildfires burning down communities were the exception, not the rule. But Cohen wanted to figure out exactly why and how these homes ignited, so he went through archived dispatch calls for burning houses and tried to pinpoint where exactly the wildfire had progressed at the time those calls were made. Homes were burning, he found, as much as a half mile ahead of the actual flames. The big flames licking up trees and into the sky weren't what was consuming the homes. The fire itself never even came to the community; residents there had by and large gotten rid of the chaparral and tall grasses in between the homes and the forest.

By the end of the 1980s, Cohen's full-time job became solving this great mystery: How do houses ignite? The term *wildland urban interface*—the area where homes intermingle with stuff that burns—made its debut in the middle part of the decade, after a particularly destructive 1985 fire season. But Cohen says that even then most others in the fire world weren't convinced that communities in the WUI were much of a concern. At a conference, a colleague came up to him and asked him why he was wasting his time studying this stuff when it just wasn't a big deal.

But in the final decades of the twentieth century, that started to change. In the early 1990s, a cavalcade of devastating fires hit developed areas in the West: the Painted Cave Fire in 1990 near Santa Barbara, where more than four hundred homes were incinerated. The 1991 "firestorm" in Oakland, which destroyed more than three thousand houses and apartment buildings. And a 1993 fire in Laguna Beach, which leveled more than three hundred homes. In the 2000s, things got even worse. Suddenly, communities were being reduced to ash. The fire problem wasn't just in forests, it was in neighborhoods too. By 2010, about thirty million homes, occupied by sixty million people, were in the WUI.

Cohen approached the issue as a physics problem: What causes fire ignition and spread? He saw scene after scene just like in the Panorama Fire. Houses caught flame miles from the head of the burning fire. Trees and shrubs were left green and standing near incinerated homes. "If the vegetation wasn't burning, then it wasn't the burning vegetation that ignited the house," he says. "And so it had to be something else." What was it?

As Cohen began to understand that the picture most people had of how homes burnt was wrong, he noted that the Forest Service and other agencies were confronting contradiction of a different sort. The agencies' desire to eliminate and suppress fire was strengthening at the same time that scientific knowledge about the ecological importance of fire was coming to light. Cohen and others found it hard to reconcile these two competing strands of thought. The government seemed to want more and less fire at the same time.

More fires burnt communities, and Cohen studied the ruins. The puzzle pieces came together. The cause of the devastation, he found, almost always wasn't the wall of flames emerging from the forest. It was what he calls "the tsunami of firebrands"—or, sometimes, low-intensity ground fire that the high-intensity flame front generates. Those firebrands, or embers, lofting through the air, can travel four miles or more if conditions are right. They land on roofs, in vents, on fences, on firewood piled next to homes. When one house ignites, suddenly there's a source for more flame in the neighborhood. The radiant heat from one home can cause those next to it to go up in flames.

Cohen's finding—that it's not the flame front itself that's the cause of burning communities—sounds sort of mundane. *Of course* embers waft into the air and catch homes on fire. But according to Cohen, that simple understanding of fire has radical implications. For him, finding a solution requires defining the problem. "An appropriate definition of wildland urban disasters is a home ignition problem," he says. "And when we start approaching the problem as a home ignition problem, then we're not trying to control wildfire." Instead, Cohen says, the way to live resiliently with fire is to change *our* behavior, not that of wildfires.

That, of course, requires social and cultural change—which is a tall order. Cohen recalls sitting in a truck bed with a colleague for a break after meeting with homeowners about fire vulnerability. They were both frustrated with the pace of change. "This isn't rocket science," his buddy told him. "You're right—it's social science," Cohen said. His pal looked at him, contemplating that thought for a moment. "Oh no, we're screwed," he said.

The thing is, we're *not* screwed. Social problems don't have technical solutions. Getting where we need to go requires working together, empathy, and collective action. Finding solutions requires recognizing that building in a fire-prone landscape is a calculated risk. If you're willing to take that risk, then there are things you can do to lessen your vulnerability to flames. This means communities and homeowners themselves must take ownership of the fire problem and what to do about it.

Since the West's increasingly devastating fire seasons are in part driven by warming temperatures, anyone living in fire-prone areas needs to support policies and politicians that confront climate change head-on. There are other, tangible actions you can take as well. Those solutions come in two forms: doing the work to make your own home resilient to fire and setting the stage for your neighbors to be able to get that work done too. The latter can mean things like zoning, codes, and regulations. It would cost billions of dollars to retrofit every home in the WUI in America—so a major government investment is important too. But solutions can also mean community mobilization—days when people get together and clean up debris in yards, meetings where people share ideas, lending tools and helping hands to people who need them.

If the problem isn't only *out there* in forests, then it's in our streets and yards and homes themselves. This section focuses on solutions at home and in your communities that can make you safer and leave room for fire to play a more sustainable and ecologically sound role on the landscape when it does occur.

It's a daunting realization, but also a hopeful one. We have the power to enact the change we need.

CITIES ON FIRE

We solved the problem of burning cities a long time ago.
—STEPHEN PYNE (TO *AMERICAN SCIENTIST*)

In the upper midwestern United States, the summer of 1871 was unusually dry. Between July 4 and October 9, the city of Chicago experienced a mere inch of rain. Northeastern Wisconsin at that point was a key source of lumber for much of the region. Chicago was in a race to catch and surpass New York as the most economically important city in the country. Building was fast and furious and powered by abundant timber.

At the time, much of the land in the region was managed with slash-and-burn techniques. Large trees were harvested for buildings and railroads. Everything else—the stuff too small to market—was burned, often in place. This intentional burning dealt with the leftover, unprofitable woody debris and regenerated the soil for productive crops to be planted the following season. It's an age-old practice; slash-and-burn has been practiced for thousands of years, bringing sustenance to cultures across the globe.

But in the Midwest in early October 1871, a strong cold front pushed through the area, elevating winds. With lots of fire on the ground already and the region abnormally dry from severe drought, a wildfire kicked off near Peshtigo, Wisconsin. It quickly exploded into a massive conflagration, torching nearly 1.5 million acres and killing somewhere between 1,500 and 2,500 people. It remains the deadliest wildfire in recorded history, yet it's not a fire that many Americans have even heard of.

That's in part because on that same day, about 250 miles south and fueled by the same conditions, something caught fire in Chicago. It's unclear to this day exactly what caused the fire. There's some lore about a barn and a cow knocking over a lantern. Some folks think it wasn't the cow but a group of men illegally gambling in that barn, or that a meteor shower caused the hay in the barn to catch fire. Other theories, perhaps most plausibly, posit that several small fires came together to set off a catastrophic blaze. Whatever the cause, nearly four square miles of Chicago burnt, fueled by wooden buildings, wooden sidewalks, and wooden roads, all dried to kindling in the drought. The fire raged for two days. It jumped across the river and knocked out the city's waterworks, leaving firefighters helpless. It basically burnt unchecked until it ran out of things to burn.

This wasn't the first time Chicago experienced a major burn, and it wouldn't be the last. Just three years later, in 1874, in the midst of rebuilding, over eight hundred buildings across sixty acres burnt to the ground. The name Arthur Ducat is one somewhat lost to history. He was an Irish immigrant whose work in the railroad industry landed him in Chicago in the mid-1850s. He entered the insurance business in 1856, one year prior to another enormous burn in Chicago, this one in 1857. He began calling for the city to get more serious about fire. He organized volunteer fire departments and demanded Chicago adopt more fire-resistant building codes and professionalize its fledgling firefighting operations. His pleas were roundly dismissed. No one liked Ducat.

During the Civil War, he served in the Union Army, rising to the rank of colonel and playing a prominent role in several key battles. Upon return to Chicago, he resumed his advocacy for smarter fire policies. In the aftermath of these enormous burns, ideas about changing how the city built gained traction. The time was ripe for reform. Building codes were rewritten, requiring fireproof materials like marble, stone, brick, clay, and limestone. Masonry emerged as a lucrative career, and an architectural revival took hold, making Chicago home to some of the world's first skyscrapers. Urban planning and zoning policies were put in place, and the city considered fire risk when making development decisions.

"Fireproof" building became a badge of honor. The rebuilt Palmer House positioned itself as the "world's first fireproof hotel." Terra-cotta and steel became centerpieces of the emergent Chicago School, an architectural style that spread to cities all over the country. The introduction of steel beams and girders enabled construction of skyscrapers with large glass windows, an innovation that combined safety, aesthetic appeal, and scale. These landmark buildings drew admiration and imitation around the world.

Of course these policies and practices didn't completely end fire in Chicago, but the Great Fire was a catalyst that raised the salience of fire-safe construction and zoning in cities across the U.S. Yes, those fancy new skyscrapers could be more resistant to fire, but older construction and infrastructure revealed key blind spots in city planning across the U.S. in subsequent years. A terrible fire in Boston in 1872 started in a warehouse basement and spread to many buildings previously thought to be fireproof. The city recognized its inadequate water system and inspection procedures. The 1906 San Francisco earthquake set fire to thousands of wooden structures. Ninety percent of the damage from the quake came from fire, and the city took notice. Another fire in Chicago, this time in the Iroquois Theater, where panicked theatergoers found blocked and locked exits and couldn't escape, resulted in 602 deaths. Fires ravaged Baltimore, New York, Cleveland, and other cities. Collectively, leaders throughout the country recognized that building was outpacing policy. Zoning policies mandating more space between buildings, strict codes regulating exits, stairwells, and construction, and innovations like automatic sprinkler systems grew ubiquitous in the early twentieth century.

Taken together, these policies and practices ended the threat of catastrophic, neighborhood-destroying blazes. Buildings could occasionally catch fire, but those fires could be contained. It wasn't necessarily easy or linear, but our cities adapted to build resilience to fire.

Much of the American West now faces a similar moment. The status quo cannot hold. There are pathways to a wildfire-adapted society, but the way forward requires something from all of us. And as with many of our most pernicious problems, making progress is more about social science than physical science. In other words, we have the tools, we just need to use them.

YOUR HOUSE

THE HOME IGNITION ZONE

Take a moment to go outside and walk around your home. What do you see? If you're like most people, you see some things that make you happy and some things that make you feel guilty: maybe trees that have flourished since you've moved in or the chores and projects you've neglected. It's easy to focus on form and function, the paint that needs refreshing or the screen door that sticks. Now shift your attention and look for things that could burn. Are there trees and shrubs close to the roof? Do you see any leaves or pine needles hanging out in nooks and crannies on the roof? How about your gutters? Any flammable debris in those? If you have a deck, what's underneath it?

Your home isn't just a place to live. It's also a potential fuel source for wildfire. It's worth considering the weak spots where it could ignite, and how flames might spread to the rest of the structure.

To get a handle on making your property more resilient, wildland fire professionals have split up the landscape around homes into a few useful categories.

The **home ignition zone** is the one hundred feet surrounding a home. This is the area immediately around your home where it can best be defended.

The first five feet out from your house is the **immediate zone**. Don't keep any flammable materials there, and make sure any plants, mulch, or other landscaping is fire resilient. As a homeowner in the WUI, this is a great place to start making change. Here, many improvements are easy and inexpensive: keep your roof free of debris and clean your gutters. Keep vents clear of debris, and install metal mesh screening to prevent embers from making their way in. It also helps to make repairs to shingles, siding, and any other weak spots that embers could penetrate. Beyond the walls of your house, remove flammable material like scrap wood and firewood.

Don't overlook anything you've stored underneath your deck. It's probably best not to have shrubs or trees in this area, and if you have wood mulch, consider getting rid of it or replacing it with a fireproof alternative, like rocks. Get to know the plants you do have better; if they have oils, resins, or waxes that could spread fire, remove or replace them. Trim any tree limbs that stray too close to your house. Focus especially on any possible fuel near the corners of your house. There, where your siding forms a ninety-degree angle, temperatures from fires tend to be higher and winds can create exceptionally tall flames. The majority of these measures are cheap; they just take time, effort, and vigilance. Make inspecting this zone of your home part of your regular maintenance. Establish a checklist to help create the habit and stick to it.

The **intermediate zone** extends six to thirty feet from your house. Here, pay special attention to the layout of trees and other landscaping in your yard. This is an area that could funnel fire straight to your house, so make sure you leave at least eighteen feet of space from one tree canopy to the next. This might mean letting go of some of the beauty and privacy you've grown attached to, but making

your yard less crowded helps create fuel breaks and makes it less likely for fire to hop from one tree to the next. Make sure the canopies of trees are at least ten feet away from any structure on your property. Also be aware of vegetation that can "ladder up" flames or serve as fuel from the ground to the treetops. Lop off lower limbs on trees you leave standing to at least six feet and cut your grass to four inches high or less.

The **extended zone** runs from thirty to one hundred feet away from your home. You're not trying to prevent a fire in this zone; you're trying to slow it down. Focus on creating gaps in fuel sources that might impede a fire's progress. Also think about keeping flames smaller and close to the ground. is means removing deadfall and unneeded shrubs that could help flames travel closer to your house. Space out larger trees, keeping twelve feet between canopies for trees thirty to sixty feet from your home and at least six feet for trees sixty to one hundred feet out. Note that the slope of your land plays a role, too, since fire moves faster uphill. Speaking with a forestry professional is always a good idea. As in the intermediate zone, keep the grass cut at four inches or less.

HIZ CHECKLIST

Zone	First Priority	Second Priority	Third Priority
Immediate (0'–5')	☐ Remove flammable debris and vegetation like dead leaves on the ground and shrubs and trees closer than 5' from the home. ☐ Remove mulch, wood piles, and scrap wood. ☐ Trim tree limbs: nothing should be in contact with the structure.	☐ Hardscape with stone, gravel, or cement. ☐ Keep any ignitable machines—ATVs, lawn-mowers, any other gas-powered stuff—out of this area.	☐ Trim overhanging trees from intermediate zone. ☐ Reconsider any attached storage structures. Move farther from the home if possible.
Intermediate (5'–30')	☐ Clear 10' around propane tanks. ☐ Cut and keep grass to 4" or less. ☐ Remove ladder fuels: anything that can transport fire from the ground to the treetops.	☐ Cut down trees. Make sure crowns are spaced out more than 18'. ☐ Avoid clusters of trees and shrubs.	☐ Landscape to create fuel breaks between things that could burn. ☐ Keep trees at least 10' from structures like fences or outbuildings. ☐ Bury that propane tank if you can.

HIZ CHECKLIST

Zone	First Priority	Second Priority	Third Priority
Extended (30'–100')	☐ Avoid/ remove piles of flammable debris: slash piles, fire wood, scrap wood, etc. ☐ Keep outbuildings clear of fuels.	☐ Remove small trees below canopy. ☐ Create fuel gaps.	☐ 12' between trees 30'–60' from home. ☐ 6' between trees 60'–100' from home.

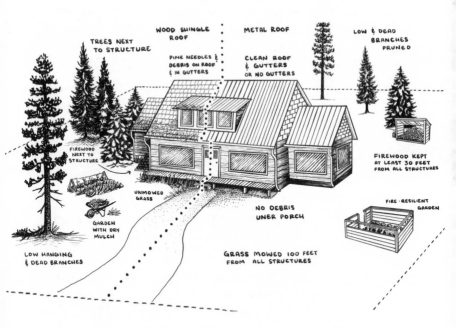

There are a number of actions homeowners can take to reduce their vulnerability to wildfires. The left side of the house above shows a house with no work done to reduce fire risk. The right side depicts a home that's taken basic actions to reduce the chances it's lost in a wildfire.

HOME HARDENING

*There's not a house out there that needs to burn
down if they just did the things that they need
to do to keep their structures from burning down.*
—TIM SEXTON, RETIRED FOREST SERVICE FIREFIGHTER,
FIRE ECOLOGIST, PROGRAM MANAGER, AND RESEARCHER

Ponderosa pine is a classic conifer of the western United States. It grows in dry climates and likes some open space. If you don't recognize it by sight, put your nose to its bark. You might be able to identify it by its telltale butterscotch aroma. The ponderosa is a great example of fire adaptation. By age five, it develops thick bark to defend itself against low-intensity, periodic fire. As it gets taller, it sheds its lower limbs, lessening the likelihood that a ground fire could ladder up into the forest canopy.

Much like the ponderosa pine, your home can never be fire-proof, but it can get pretty close. It can become resilient to most fire risks as they approach. And there is often help available. Many WUI communities have grant programs to fund wildfire risk assessments and support investments in hardening your home. If you are considering renovations or new construction, there are three important categories to think about.

First is roofing and vents. One of the main threats of a wild-fire burning your home is floating embers, so your roof represents the largest surface area on your home available to those embers. About one million homes in medium-to-high wildfire risk areas have wooden roofs. These can catch fire in an instant. Headwaters Economics estimated it would cost $6 billion to replace all roofs in the WUI in the country with fire-resistant materials. We're not holding our breath for the federal government to pay for all this, but if you have to replace your roof, look for what are called "Class A" materials. These include metal, composite shingles, concrete, and clay tile. Remember that a fancy new roof won't do you any good if it's not clean. Don't let flammable stuff gather up there, and check it regularly. If getting on your roof is not something you're able to do or comfortable doing, ask a neighbor or hire a professional. Most gutter cleaning companies will blow debris off your roof while workers are up there.

Attic and crawl space vents are another common weak spot that embers can exploit, a problem compounded by the things we tend to store in those places—cardboard boxes, files, and lots of other highly flammable stuff. These vents, however, are critical for managing moisture and airflow in nearly every home. Eliminating vents completely is possible with some modern building techniques, but that's really only an option for new construction. The best solution is to screen the vents. Most building codes and home inspectors address this to an extent, but the National Fire

Ponderosa pines are an iconic tree in the West. Historically, they naturally burnt every five to thirty years, depending on their ecosystem. But fire suppression put a stop to that natural cycle, increasing the fuel load in many forests.

Prevention Association recommends a one-eighth-inch screening as the best balance of fire and airflow safety. Putting up these screens is relatively easy and inexpensive—all you need is the screening material, tin snips, a staple gun, and a hammer. When you get to installing, study your home carefully to make sure you don't miss any vents, like those you might have under your eaves. An eave is the edge of a roof that overhangs the side of a house. Sometimes eaves are open and sometimes they're connected to the side of the house with what's called a soffit. Studies have shown that closed or soffited eaves are more ember-resistant than open eaves, so close them off if you can.

The next category to think about is stuff that might be attached to your house—decks, fences, gutters, and the like. Like roofs, decks present another large landing area for floating embers. If you have a deck, never store firewood, scrap wood, wooden rakes, or anything else flammable underneath it. Keep that space underneath clean of flammable debris as well and grab a screwdriver or putty knife and clean out debris from between the deck boards too. If you're building a new deck, consider a nonflammable material, like a Class-A composite wood or aerated concrete. These are more expensive, but generally have a longer life. Space out the boards one quarter inch, as opposed to the typical one-eighth-inch gap. Spread the joists out farther too—ideally twenty-four inches. That allows debris to fall through the deck, rather than gather in those tiny gaps. And whether your deck is new or old, make sure the corners where it attaches to your house are clear of debris. This is an easy spot for leaves and other fuel to accumulate. When it comes to fences, think of them as fire delivery devices. If a fence catches fire and it's attached to your house, it'll deliver those flames straight to your living space. Replacing the whole fence could be costly—but one solution could mean building with similarly nonflammable composite or concrete materials within just five feet of any building. Use materials and designs that tend to capture the least amount of debris. These often tend to provide the most airflow. And remember that even a nonflammable, metal fence can be a problem if the areas around it capture leaves, pine needles, and other flammable debris.

Gutters are the final threat in this group of stuff attached to your home. If you're designing a new home or renovating the one you have, gutters can often be eliminated. If you need them, however, keep them clean. Gutter screens can help, but they often get clogged up. A simple blower is the best tool for the job. Get up there and blow out everything that's collected. Be careful on your roof or up on a ladder—ask for help or hire it out if you need to.

The third category to consider is siding, windows, and trim materials. This stuff is important, but the other areas—roofs, decks, gutters, vents, debris—should be addressed first. Siding and trim can be noncombustible, ignition-resistant, or combustible. Noncombustible materials are the gold standard. They won't burn and include things like fiber-cement, metal siding, and three-coat stucco. Ignition-resistant materials, like composite wood, brick, and stone, will withstand most conditions, but can still ignite under extreme heat. And with stone or brick siding, note that while they do a nice job protecting your home, they generally provide only a thin layer sitting over wooden sheeting. In extreme heat, the wood underneath can ignite.

Wood, compressed wood, and vinyl siding are all combustible and should be avoided in new construction and renovations. Some woods are treated with fire retardant and are rated Class A. They are less expensive but offer less protection. You don't necessarily need to ditch your combustible siding now. If you're beginning to invest in hardening your home, start with the other stuff. But when it comes time to replace that siding, consider a material that won't burn. Windows are another weakness. Yes, glass doesn't burn, but it can break when stressed by heat, offering free passage for embers and flames. Your windows should be dual-pane, tempered glass, which does a good job resisting heat stress.

Beyond what your home is made of, think about the overall design too. Some vulnerabilities are less about material weakness and more a consequence of design choices. Complicated architectural designs with intersecting roofs and parapet walls are more vulnerable because they create more surface area for embers and flames to attack. They also create heat traps, spaces of low air flow that increase the likelihood of an ignition. If you can see a spot on your home where water could collect and puddle, that's a spot where debris can gather too. It's also a spot where a burning ember

HOME HARDENING CHECKLIST

Zone	First Priority	Second Priority	Third Priority
Roofing & Vents	☐ Clear all flammable debris. ☐ Clean gutters. ☐ Repeat regularly! ☐ Trim tree limbs within 10' of chimneys or overhanging roofs.	☐ Add screens to eave, soffit, and crawl space vents with ⅛-inch mesh. ☐ Weather-seal garages.	☐ Use Class A roofing material ☐ Skylights should be glass, not plastic.
Attached Structures	☐ Keep clear of debris. ☐ Clean between deck boards. ☐ Don't store anything flammable on or under your deck (remove immediately if fire is near).	☐ Use nonflammable fencing within 5' of house. ☐ Use non-combustible deck materials	☐ Space deck boards ¼ inch. ☐ Spread joists 2' apart.
Siding, Windows & Trim	☐ Check siding for damage, cracking, or rot. ☐ Plug gaps greater than ⅛ inch. ☐ Repair any broken windows or screens.	☐ When it's time to upgrade, install dual-pane, tempered glass windows and sliding doors. ☐ Install 6" or longer flashing between siding and any decks.	☐ When it's time to remodel, use fire-resistant design/ architecture. ☐ When it's time to remodel, install non-combustible siding.

can linger. And if that burning ember happens to land on something flammable, you'll likely have a home ignition.

Do your best to pay attention to everything surrounding your home: from the raggedy junipers sprouting near your siding to the patches of dense sagebrush farther away from your door. "It's the little things that are important," Cohen says.

In many WUI communities, there are dedicated wildfire professionals who can help you assess your risk and figure out what resources are available to help you and where to start. Many places offer cost-share programs or financial incentives to get this work done. Reach out to your city, county, and local Forest Service office, ask neighbors, even just try googling—find out what's available in your neck of the woods.

WHAT DOES RISK LOOK LIKE?

For many folks living in the West, wildfire risk is synonymous with mountains, canyons, and pine trees. But real risk can come when and from where you least expect it.

Take, for example, the Marshall Fire. It erupted in late December 2021 just outside Denver, Colorado. It destroyed over one thousand structures in the towns of Louisville and Superior—including entire subdivisions, commercial buildings lining a six-lane highway, and a hotel. The fire roared up no mountains and burnt few trees. Instead, high winds blasted it through flat, high plains.

The conditions were especially ripe for a devastating burn. A moist spring and early summer allowed grasses to grow tall and dense. Then, a bone-dry second half of the year and sparse early winter snow left the landscape parched. A lawsuit alleges a downed power line ignited the blaze, which rapidly moved to structures. Driven by winds of over one hundred miles per hour, the fire blew from house to house. Twenty-four hours later it was mostly over, the smoldering foundations blanketed by an eerie coat of fresh snow. It looked like a nuclear winter.

GO-BAG

If you live anywhere near fire-prone land, an emergency go-bag is essential. Don't wait for a fire to sprout nearby to start packing one. Begin now. Some key items for a go-bag can include a sturdy pair of shoes, leather work gloves, goggles, N95 masks, a map, prescription medications, water bottles, energy bars, spare phone battery, keys and credit cards, a first aid kit, headlamp, a spare toothbrush and toothpaste, personal hygiene items, and pet supplies. Add the things that might help keep you sane in the event of an emergency: maybe a pack of playing cards, headphones, a favorite book.

Go-bags need annual inspection and renewal. Dump expired prescriptions and replace them with fresh ones; dump expired food and replenish with new food for yourself (and for your animals, if you've got them). If there are any updates to your essential documents (advance directive, insurance, bank account, other emergency numbers, etc.), make sure those get changed out. When the bag is ready to roll, keep it somewhere easy to access when you need to make a quick exit.

As you put together your go-bags, include your children and loved ones in the process. The process doesn't have to be ominous, threatening, or scary. Making it a family event can add a flair of fun and promote a sense of togetherness as everyone prepares for an evacuation.

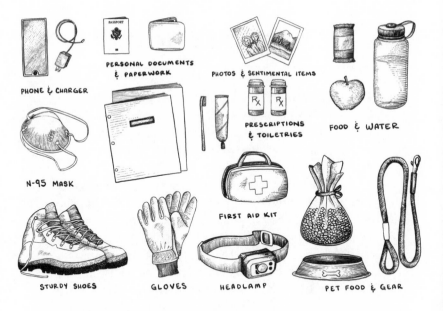

PHONE & CHARGER

PERSONAL DOCUMENTS
& PAPERWORK

PHOTOS & SENTIMENTAL ITEMS

PRESCRIPTIONS
& TOILETRIES

FOOD & WATER

N-95 MASK

FIRST AID KIT

STURDY SHOES

GLOVES

HEADLAMP

PET FOOD & GEAR

Preparing go-bags, full of necessary gear to help you survive in the event of an evacuation, can help you and your household prepare for a wildfire and other emergencies.

A similar story played out in Talent, Oregon, during the 2020 Almeda Fire. The burn started in a creek bed and moved north through the riparian zone, consuming the abundant grasses and cottonwoods. The creek parallels I-5, the major highway in the area. Soon fire reached the south end of Talent. It quickly found homes—primarily mobile homes—and burnt most to the ground in moments. About seven hundred buildings were destroyed.

Fires in the plains and valleys, along creeks and rivers, isn't the scenario most folks imagine when they think about wildfire risk in the WUI. But more and more, the reality of wildfire includes places like Talent and the suburbs of Denver. Densely constructed homes offer lots of opportunities for wind-driven embers to find weak spots—some debris on a roof, an uncleaned gutter, a propane tank, a rooftop terrace, a townhome under construction. Once

RED FLAG DAY

Red Flag Warnings are issued by the National Weather Service. They indicate an especially dangerous combination of high temperatures, low humidity, and high winds, creating extreme wildfire risk and the likelihood of particularly dangerous fire behavior.

When you see a Red Flag Warning, take extra special care with anything that could spark or burn. That should obviously mean no campfires or planned burning, but also be mindful of anything you do that could cause an errant spark. Maybe it's not the best day to run that chainsaw or lawn mower. If there's something going on with your car, say, it's backfiring or dragging something that sparks, don't drive it on a Red Flag day (and when the warning is over, get that thing fixed!). If you like hitting the trails with your motorbike or side-by-side, take Red Flag days off.

If there's a fire nearby, Red Flag Warnings also might indicate that fire activity is likely to grow. Things could get a lot worse, and fast. Fire managers typically put more folks on duty during these times, have more engines on hand, and keep fires staffed twenty-four hours a day. This level of vigilance, however, is hard to maintain when Red Flag Warnings string together over multiple days or span a broad geography. Keep an extra-attentive eye on fire behavior nearby through InciWeb, and your county and city communications, on days like these.

the fire finds a structure, the nearby structures become the primary source of fuel. Now the fire becomes a threat to the entire neighborhood. Wildland firefighting tactics—digging line, backburns—just won't work or are impossible to safely execute. Urban firefighting systems, like fire engines, ladder trucks, and pumpers, are overwhelmed by the sheer number of homes to defend.

Prior to the Marshall Fire, the Forest Service rated the area as high risk for wildfire. But what a place looks and feels like might cloud our judgments of this risk. Don't take your home's safety for granted. Whatever you consider your community's risk to be, check that perception against what the experts think. Visit wildfirerisk.org, a website developed by the Forest Service with interactive risk maps and loads of other information.

EVACUATION NOTICES

When a wildfire is approaching, you need to know what's going on, especially if you need to grab that go-bag and evacuate. The flames can knock out cell towers, telephone lines, cable television, and radio signals. It's unlikely, however, that all those communication channels will go out at once. Sometimes, evacuations can be necessary in a matter of minutes, or even seconds.

So it's critical to sign up for emergency notifications through multiple outlets. The Federal Emergency Management Agency (FEMA) has established the Wireless Emergency Alert system that sends short messages with critical information to every compatible mobile device in the country. Check with your wireless carrier to make sure your device is compatible. The WEA system is designed for similar situations, but takes over all broadcast media—television, radio, and so forth. Emergency notifications are also pushed out through the NOAA National Weather Service; you'll see them in your weather app. FEMA also has a mobile app, and if you live in a wildfire-prone area, you should definitely download it.

A big challenge, however, is that wildfire threat is typically local, and local emergency and evacuation notification systems vary widely. Search for "[your county] emergency alert system," and you should find directions for enrolling. Many counties in the U.S. use the Smart911.com platform, which targets emergency information to specific geographic areas. Landlines are automatically enrolled in this system, but mobile phones are not. Enroll your mobile phone number now.

Many areas prone to wildfire have spotty cell phone coverage. Service will likely deteriorate or fail during a wildfire, so you might consider an emergency satellite communicator, like a Garmin inReach or a SPOT device. Practice with the device and know how to use it. Preload the contact numbers of critical friends and family so that you can message them in a hurry if you need to. Increasingly, smartwatches and smartphones are available with emergency sirens and satellite beacons.

The technology will improve, but it can't necessarily fix everything. Be proactive, find out how to enroll in your local area, and enroll. Then, tell your friends and family. Or even better, enroll them yourself!

FIRE-RESISTANT PLANTS AND GARDENING

Take a look at the plants around your home. How many are native to where you live? Are there any invasives? It's likely the native plants are the healthiest and the most adapted to the area. And if you live where wildfires regularly occur, those plants likely have some kinds of adaptation to the flames.

All plants can burn, but some burn more easily than others. And more important, where you put your plants and how you care for them is critical in reducing wildfire risk to your home. There should be no flammable vegetation (and that includes wood mulch) within five feet of your house—that's the area fire folks call the "immediate zone." Farther out, it's okay to have trees and other plants, but you should consider the following questions:

- Does the plant contain a lot of oils, resins, or waxes? The higher the level of each of these, the more flammable the plant. They also burn hotter.

- Are the plant's limbs adequately spaced out and balanced, or are they packed densely together in a tangled web? Dense limbs are more likely to capture embers and catch fire. Limbs should be spaced out about 3 percent of a tree's expected height. If the tree will get to fifteen feet, you want the limbs about five and a half inches apart.

- How fast and how tall will the plant grow? The faster and taller it grows, the more maintenance and concern for other plants is required.

- How much stuff—bark, leaves, needles, etc.—does the plant shed? Everything a plant drops is fuel and needs to be cleaned up regularly.

The good news is that any species native to a wildfire-prone area will likely provide favorable answers to those questions. But there is no single list of fire-resistant plants. Researchers say such a catalog might be unreliable and potentially misleading. After all, how you care for your plants plays a much larger role than the plant species itself. Plus, ecosystems vary widely throughout the West. When deciding where to put your plants, think of islands. Avoid connected swaths of vegetation, as they provide a continuous fuel source. Islands offer fuel breaks. For larger plants, pay attention to any canopy you've created. Consider the minimum distance between any part of a plant, not just the distance between the trunks.

FIREWISE USA

The Firewise USA program was designed to bring Jack Cohen's revelations into practice. As more fire experts started focusing on the WUI, *firewise* emerged as a term to describe "doing the right thing." Since then, it's grown into a set of community and homeowner standards designed to make neighborhoods in the WUI as resilient to wildfire as possible.

The website (FirewiseUSA.org) is a valuable repository of information, with a wealth of resources for both individual homeowners and community leaders. Every state in the country has a Firewise USA liaison whose contact information is publicly available. These folks serve not only as points of contact for the public but also as extension agents for the various federal and state agencies tasked with wildfire suppression and risk mitigation. Firewise USA organizes hands-on workshops that offer actionable information on fire behavior and home safety, along with the training to use that information.

Participating in these workshops not only creates shared knowledge, it builds community. It shows community members that they're not alone in confronting the wildfire challenge and that they have agency to keep themselves and one another safe. Firewise USA isn't a one-stop solution to the wildfire problem. But it's a critical tool for building community awareness, empowering neighbors, and promoting action from the grassroots level.

Watering presents tricky trade-offs. Given the expense and scarcity of water, especially in the West, there are plenty of reasons to minimize watering. But dried-out plants burn more easily. If you choose plants that require a lot of water, the best thing to do to minimize wildfire risk is to water them sufficiently. Consider getting to know your ecosystem even better. Learn about what grows and thrives there. Your local Forest Service office, a local university extension agent, or even a local wildfire community group might have more information on this. Native plants tend to be more fire resilient and to need less water.

Bottom line: pick native plants, place them carefully, and maintain them regularly.

WHERE THERE'S FIRE, THERE'S SMOKE

Across the West, wildfire smoke is typically the most significant source of air pollution. It's increasingly common to see smoke clogging the skies across the entire country—from San Francisco to New York. During the height of fire season, major coastal cities now regularly experience air quality on par with some of the most polluted places on earth.

Fortunately, there are simple and inexpensive things you can do to mitigate the negative effects of wildfire smoke. First, understand the risk. Smoke is bad for all of us, but it's particularly dangerous for pregnant people, people with respiratory conditions, children, older folks, and other vulnerable groups. Folks who live in older homes or work in buildings without modernized air filtration systems are especially at risk. Many people work outside, and lots of others just love spending time out in the woods. After all, recreational time in the outdoors can be a critical part of mental and physical health. The more time you spend in the smoke, the worse the effects.

Second, know how to check the air quality. State and local health departments publish air quality ratings on their websites. Bookmark these sites and sign up for their reports. The specifics

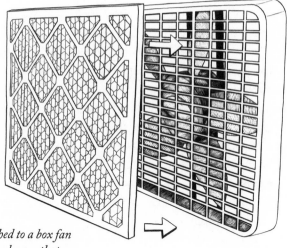

A simple filter attached to a box fan can create a DIY air cleaner that can increase the comfort and livability of your home when the skies are smoky.

depend on where you live, but there are smartphone apps, like AirNow, AirVisual, and other weather apps, that aggregate these ratings. More options are certainly in the works as well. Home air quality monitors are improving, but the good ones are still expensive.

Next, make sure the air in your house is clean and limit your time outside. There are many excellent air purifiers on the market, but make sure whatever you get has an authentic HEPA designation and check that it blocks particles down to .3 microns. Anything larger than that will allow wildfire smoke through. Also pay attention to the room size for which the purifier is rated. These machines can get expensive, but many communities and public utilities have grants and rebate programs to help folks get the protection they need.

At the same time, do what you can to make sure your home isn't leaky. Make sure windows and storm windows are closed. Replacing old windows and sealing in gaps can be a big help. Some of this work can go a long way toward energy conservation as well. Avoid cooking on a gas stove inside; try to use the oven or microwave instead. Also avoid vacuuming and any other house-hold chores that might get settled smoke particles flying around.

Beyond these individual actions, check in with your neighbors, particularly the older folks, to make sure that they're doing the same stuff. For those who need help, a standard air conditioner filter (make sure it's HEPA!) affixed to a cheap box fan is a great stop-gap measure. Make a few and give those to people in need. Many communities have resources to help you find help if you need it. And if you can offer some assistance, support these groups and the people who need their support.

FINDING THE TIME: A GUIDE

Improving at anything is hard, whether you're getting fit, learning a new language, or becoming a better partner. These things all take time and energy, two things in scarce supply for most of us. Home improvement is the same. It can feel like a chore, and the myriad other demands of daily life can take priority. So maybe you feel overwhelmed by what you've read so far. All these new things to do. You might think: *More effort, time, money that I don't have. Ack!*

But it doesn't have to be that way. There are lots of ways to split up and segment the work to make it feel much more manageable. To get started, identify your goal. Maybe it's knowledge and understanding. Maybe it's protecting your home and loved ones or becoming an effective community leader. Specify your goals and how important each one is to you. If you live with other people, include them in the conversation. Get on the same page about what's most important to you and what you want to achieve.

Next, take stock of where you are now. If you own a home in a wildfire-prone area, get an expert assessment. That process will identify the big picture of what needs to be done and in what order.

The list below is another way to split up the tasks above into easier categories, which could help you get more done—even if you've just got a few minutes during your lunch break.

Low-Hanging Fruit
When you've got a few extra minutes

- Make sure you're enrolled in your area's evacuation notices.

- Call a couple of neighbors. Check in about evacuation plans and home-hardening projects. Offer to lend tools or your own time if available.

- Build a DIY air filter to prepare for smoky days.

- Find out if there are any local fire-readiness groups around. If not, begin to organize one on a day off.

- Call the local Forest Service to learn more about how the agency is preparing for fire around where you live, along with any fuels treatments and prescribed burn projects in the works.

- Prepare for Red Flag days.

- Talk with your family about evacuation plans. Where will you meet? Develop a plan for your pets.

After Work
*Tackle one of these items at a time. Try making a dent in the list—
one item a week, or even a month.*

- Replace a vent with ⅛-inch screening.

- Clean out under your deck.

- Put together a go-bag.

- Find out if your city or town has a Wildfire Preparedness Plan; find the nearest Firewise USA site.

- File a public comment on any proposed forest projects near you.

- Attend a listening session for a forest project proposal.

On Your Day Off

- Clean your gutters.

- Move your firewood and scrap wood away from the house.

- Limb and prune your trees and shrubs.

- Cut down some trees.

- Get rid of the wood mulch within five to ten feet of your house. Replace it with gravel or another noncombustible material.

- Get rid of other vegetation within five feet of your house. Try to keep that small area as fire-resistant as possible—meaning no sources of fuel.

- Volunteer to help a neighbor get some important work done.

Big Projects

These will take from days to weeks and will require a relatively large investment. There might be resources available locally to help you fund these projects, and that sort of assistance is growing.

- Get a new roof. Fire-resistant materials include asphalt shingles, metal, or concrete, brick, or masonry tiles.

- Build a fire-resistant deck.

- Replace your fence with fire-resistant materials.

- Install fire-resistant siding.

- Get rid of mulch next to your house and relandscape with fire-resistant plants.

- Organize your neighborhood and build community awareness of the home ignition zone.

A FIREPROOF HOUSE?

Given the shortage of housing, continued migration to the WUI, and climate change, there is increasing demand for disaster-proof home designs.

Architects and builders are meeting this market opportunity. Metal roofs and noncombustible siding are one thing, but designs focused on eliminating fire risk from the get-go are relatively new. In Paradise, California, what look like enormous, half-buried soup cans are sprouting up on some lots where more traditional Craftsman and ranch-style homes stood before the 2018 Camp Fire. These glorified Quonset huts are based on the temporary, shippable, quick-to-assemble storage depots used by the U.S. Navy in World War II. They are sturdy, mostly steel, and commonly used throughout the country as outbuildings.

This Quonset hut design eliminates many of the wildfire vulnerabilities of more traditional homes. There are no corners or crevasses where embers can find and ignite dead leaves and pine needles. The aluminum window frames won't melt under intense heat as quickly as vinyl frames. There are no overhangs and minimal venting, classic weak spots in other home designs. Everything exposed to embers and flames is noncombustible metal or concrete. Inside, however, these structures can look and feel like modern homes, typically utilizing a contemporary open-concept design with no barriers between the kitchen, living space, and dining room.

As communities enact stricter building codes and development in the WUI continues, we're likely to see more design innovations like this. Outside of a completely concrete bunker, however, no house is completely fireproof.

YOUR COMMUNITY

IT TAKES A VILLAGE

The fixes required to create defensible spaces around homes and to make buildings themselves more fire resilient are relatively low-tech and easy: get rid of dense trees, shrubs, and other flammable foliage, replace flammable roofs with things like metal alternatives, make sure all openings are covered up by one-eighth-inch mesh and that no debris gathers in gutters. Making changes like these are resource intensive but don't require fundamental shifts in the way we live. But zoom out, away from individual homes, and focus on subdivisions, neighborhoods, entire communities. Suddenly, there's a twist.

All over the West, there are dense developments. In an instant, that individual problem becomes a collective one. That one-hundred-foot perimeter of homes that creates the so-called home ignition zone? In those neighborhoods, there can be four to six houses within a single HIZ, says Jack Cohen. As a home-owner, the problem isn't just yours anymore. It belongs to the entire community. One bad actor in a community can create a hazard for everyone else when fire comes to town. "It changes the social dynamics to where we as a community have to cooperate in increasing each home's ignition resistance in order to create an ignition resistant community," Cohen says.

When one home catches fire, it starts to create its own heat and embers. Wind blows and distributes those embers from the house, rather than the wildfire itself, to other homes in the area. Suddenly, it isn't the forest that's on fire. It's the community. Fire spreads house to house.

This means homeowners—and entire communities—need to take collective responsibility for the wildfire problem. "The home should not be considered a victim of the wildland fire," Cohen writes, "but rather a potential participant in the continuation of wildland fire."

You could clean your gutters every day, but if your neighbor's house goes up in flames, yours will probably go too. Studies of communities ravaged by fire confirm this. And so do accounts from firefighters on the ground in WUI neighborhoods. They typically see fires spot from home to home, often skipping the nearby vegetation altogether. The fire will find the weakest link and the best source of fuel.

That means the focus can't just be on fire-adapted *homes*. It needs to be on fire-adapted *communities*. Doing so relies on neighbors working together to mitigate as many wildfire risks as possible, creating a whole much larger than the sum of its parts.

The reality of the effort required to reach that goal creates what social scientists call a "collective action problem." Those problems permeate many of society's trickiest-to-solve challenges: vaccination, climate change, and voting—just to name a few. A collective action problem, basically, means that every single community member will be better off if everybody participates. But with so many participating, some might feel like they don't need to act at all; they "free ride" off the efforts of others. In the case of wildfire, that might mean certain members of a neighborhood might shrug off the necessity of doing the work on their home that's necessary. They might look around and think: well, my neighbors have already done it, so I oughta be pretty safe. That reasoning, though, is fatally flawed. A central challenge in reaching the scale of building fire-safe communities required is getting everybody on board.

This, of course, is hard to do. Many homeowners simply expect the Forest Service or local fire department to come to the rescue whenever they need it. A century of fire suppression built this expectation; people think managing wildfire is someone else's job, a service the government provides. Beyond that, many folks have developed a sense of "learned helplessness"—since someone else has always dealt with wildfire, many have never learned a thing about it.

There's no single fix to all these barriers. However, that doesn't mean the problem is insurmountable. If communities across the West can succeed in the monumental effort ahead, it will have enormous implications for our future with fire.

For one, shifting our collective focus to making communities more fire resilient can mean fire management agencies have more room to let fire play its natural role on the landscape. They'll have the leeway required to shift strategy away from full suppression. That would initiate the kind of feedback loop our forests and people need. As fire scars prevent the spread of new burns and ecosystems return to health, that will in turn make communities even safer from out-of-control burns in the future.

Achieving this future means reformulating our relationship with fire, and with one another.

BUILDING TRUST

Here are some depressing data: Americans' trust in major institutions like the three branches of government, newspapers, the criminal justice system, and the police are at their lowest levels since the polling institution, Gallup, began tracking trust about fifty years ago. Trust in our neighbors has eroded as well, especially among younger people.

This single factor—trust—is a backbone to a thriving democracy. But it also has huge implications when it comes to wildfire. People are reluctant to accept the expertise of local, state, and federal resources and guidelines that can help make buildings more fire resilient. They're less likely to work together to build fire-adapted communities.

This is where Libby Metcalf and Alex Metcalf come in. They're social scientists at the University of Montana's College of

Forestry and Conservation. Most of their colleagues study how trees, rivers, and entire ecosystems function, but these two study how *people* operate. They home in on how community members make decisions and develop attitudes about the ecosystems in which they live. In particular, the Metcalfs zero in on the often fraught relationship between community members and federal agencies like the Forest Service.

That relationship can influence both forest projects—like prescribed burning and thinning—and how much mitigation work gets done on homes and properties. According to the Metcalfs, most official efforts to encourage more work in the WUI have hinged on an "information deficit model." That model suggests that if people just *knew* more, they'd act differently. So in the case of fire resilience, it makes a big gamble: if agencies can feed people the right nuggets of information—as in a lot of what we've detailed above, in terms of the work that needs to get done on individual homes—they'll do that work and make themselves and their communities a lot safer. However, the reality of the human decision-making process is a lot more complicated than the information-deficit model assumes. Humans have complex desires and values—and getting people to act requires addressing many factors that lie beneath the surface.

The Metcalfs interview community members, conduct meetings, and give out questionnaires. In social science terms, they employ a mixed methodology—looking quantitatively at what people have to say but also having deeper discussions about what really matters to them. They've observed communities navigating complicated and divisive topics, but the heart of the matter is often quite simple. So much of what people desire, and what informs their decision-making, comes back to a basic component of American existence: life, liberty, and the pursuit of happiness.

Just like most folks, people living in wildfire-prone lands want their lives, families, and homes protected. They want to make sure that they can prosper even as the landscape around them is changing. This, the Metcalfs argue, is where government agency efforts to make communities more fire resilient should focus. But doing so requires getting communities and agencies to see eye to eye, which is a tough task.

Historically, the forestry profession selected for folks more interested in trees than people. But today, forestry isn't separate from communities across the West. It affects landscapes in which people live, work, and play. So community members, the Metcalfs argue, need to be an active part of any decision-making process. They need to be seen, heard, and respected. The process, they contend, is much more important than the outcome.

The social science tells us that effecting change takes genuine engagement and relationship building at the local, and even individual, level. An edict from on high won't do it. The act of fireproofing communities is by necessity bottom-up. Just as each forest is different, every community is different as well. One-size-fits-all approaches alienate almost everyone and are ineffective. "Our relationship with wildfire and the way that we respond to it has to be determined by local conditions," Alex Metcalf says. "Both the ecology of it and the role of fire, but also the people and what they care about and how they choose to chart their path forward."

CASE STUDY: BOULDER COUNTY

Boulder, Colorado, has a bit of a reputation. Up against the Front Range of the Rockies, less than an hour from Denver, it's thought of as a progressive haven. The city is home to Olympians and PhDs, and it's extremely affluent; the median home price in Boulder in 2022 was nearly a million dollars. But this wealthy, liberal community is also at the forefront of wildfire risk mitigation. It's easy to assume that's due to the progressive politics and money. But the factors that influenced homeowners' decisions to fireproof their houses and lawns are worth a deeper look.

Boulder was one of the first communities in Colorado to experience wildfire. Fires in 1988, 1989, and 1990 burnt through more than forty homes and thousands of acres. A few years later, the county's fire chief proposed that the area outlaw wood-shingle roofs. The opposition was fierce; the wood industry sued both the city of Boulder and the fire chief. The fire chief had defamed the industry, the suit claimed, when he said "Wood burns." Wood does, of course, burn. So the county prevailed. They outlawed wood roofs, with provisions that made the change manageable for existing development. At the same time, the county began planning subdivisions with wildfire safety in mind and identifying fire risk countywide. But there was still a problem: thousands of people's homes weren't up to snuff when it came to wildfire.

When Jim Webster joined the county government in 2009 to help write a fire preparedness plan, he wasn't a career fire guy. He'd started out as a Peace Corps volunteer, and then a community organizer and policy analyst. But the task of fireproofing homes, he found, wasn't so much about fire. It was about organizing. He felt right at home. The need for fire mitigation measures was immediate; in 2010, the Four Mile Fire destroyed 269 homes in the county. By 2014, Webster helped start Wildfire Partners. It's a program run by the county, but it feels more like a neighborhood action group. Webster wanted to shift the modus operandi of wildfire groups. Instead of hosting community dinners and sending mailers, he figured it could be much more effective to go door-to-door and target homeowners one by one. To do it, he needed team members who could offer thoughtful advice and truly understand risk. Luckily, they were in Boulder. And he found them. "They're PhDs, former fire chiefs, former district foresters," he says.

One-on-one interaction also requires figuring out messaging. That means Wildfire Partners' staffers aren't just knocking on doors and saying "You're at risk of burning down." Webster says they tailor what they talk about based on the person: it could be pet safety, it could be family, it could be wildlife. When receptive homeowners get on board, Wildfire Partners provides them with a detailed checklist, photos, and ideas on how to begin the process. Staffers have also worked with the state and federal government to secure more resources and provide financial assistance to those that need it. The focus is not just on what needs to get done on homes but also on *why* that work is so important.

At the same time that Wildfire Partners was getting up to speed in Boulder County, another issue was arising for homeowners in wildfire-prone areas: insurance. All over the West, including in Colorado, insurers were scrambling to deal with the wildfire crisis. In Colorado, some folks reported getting letters of cancellation from their insurers. Others said that their insurers required certain mitigation measures, but the reasoning behind those measures was vague, and how to get the work done was hard to follow. Folks interested in purchasing a new house might face denial after denial before turning to an expensive policy with less than ideal coverage. Even for those not facing immediate impacts of these insurance changes, there was a palpable dread and uncertainty among homeowners in and near the WUI.

While policymakers scrambled to figure out how to handle these issues, Webster saw an opportunity locally. He figured there had to be a better way, so he started talking with insurers themselves. "I was lucky," he says. He just had to focus on Boulder

County—not an entire state. As he started conversations with insurance companies, he wasn't telling them they were *wrong*. "Our philosophy was the insurance companies are our partners," he says. "We don't want the homes destroyed, the insurance companies don't want the homes destroyed. Let's work together."

In the end, Webster was able to get many major insurance companies to agree to a deal. If homeowners received certification from Wildfire Partners for their mitigation work, then their homes would be insurable. Instead of dealing with cancellation, higher premiums, or calling around until a company agreed to give them a policy, homeowners could rest a little easier knowing that they weren't at risk of losing their policies. The deal also offered a sort of indirect discount. The name-brand insurance companies tended to be much cheaper than what Webster called "the providers of last resort," or the more expensive policies with less coverage that homeowners might turn to if they fail to qualify for more traditional coverage.

When Wildfire Partners certifies a house, homeowners receive a yard sign. That helps get word out about the program—and makes the often subtle mitigation work immediately visible. Webster says many people display those signs like a badge of honor. But the work doesn't end there. "It's not a one-time thing," Webster says. "It's not just getting a shot or showing up at the dentist. There's a lot more to it than that." Plants grow. Pine needles fall. Homeowners need to continually maintain their yards. They need to keep their gutters clean. Fire resilience is an ongoing process.

Nearly a decade into the program, Wildlife Partners had worked with thousands of homeowners in Boulder County. Word got out so much that they didn't need to recruit anymore; people came to them. But then came a plot twist. In late December 2021, the Marshall Fire hit the area, destroying more than one thousand homes. Those houses weren't among the nine-thousand-odd homes at risk of fire in the mountains. They were in the suburbs of Denver, torched by a grass fire fueled by drought and high winds. Suddenly, tens of thousands more people realized that they weren't safe from wildfire. On the one hand, the increased demand is near impossible to meet. But Webster says the Marshall Fire opened people's eyes. The wildfire problem suddenly became all too real for folks who thought they were far from it.

In nine years, Webster says Wildfire Partners has worked on nearly three thousand homes. Part of that success, he says, is because the measures they're asking folks to perform aren't mandatory; they're voluntary. "It's not coming from government," he says. "It's coming from all of us. It's like 'Hey, we're neighbors, we're here to help you.'"

NUDGES

Richard Thaler, a Nobel Prize–winning behavioral economist, says, "People aren't dumb. The world is hard." His work recognizes that people don't always act in their best interest. And when it comes to wildfire mitigation, that's an important starting point to understand why people behave the way they do.

Thaler pioneered the idea of "nudges," the notion that tiny elements of how information is presented can have big effects on the choices we make. His favorite example of a nudge is something he noticed in an Amsterdam airport: small images of flies were painted near the drain on urinals in the men's room. The airport reported an 80 percent reduction in "spillage" after those flies were added. "Men evidently like to aim at targets," Thaler told the *New York Times*. Some classic examples of "nudges" include things like automatically checked boxes—as in enrolling employees in a 401(k) by default instead of asking them to opt in—utility provider mailers that compare one household's energy consumption with that of their neighbors', even activities as mundane as sending people reminders to schedule their next doctor's appointment.

While nudges have been harnessed in finance, education, and even fine dining, the field remains relatively young in wildfire mitigation research. But there's tremendous promise: if basic nudges

A NEW NORMAL

Neighborhood norms can be a powerful force influencing wildfire risk mitigation.

Look around your neighborhood, and you might be able to pick out a multitude. How often do your neighbors mow their lawns? How long after snow do they shovel? Norms like these are things we often take for granted, but they play a huge role in our daily lives. The Metcalfs recounted their own experience with a norm: they moved into a new house and were discussing what type of trash can to get. The house was right on the edge of where black bears might rummage around during certain times of the year. They noticed that all their neighbors had bear-proof trash cans, so they decided to get one too.

In terms of wildfire, if taking steps to mitigate household wildfire risk in a particular community is what most people do, that creates pressure for newcomers or holdouts to take the same action. People don't want to be left out or left behind. Most actions you can take to lower your wildfire risk are visible to your neighbors, be it clearing flammable debris from around your home or installing a new roof. Neighbors pay attention, notice things, and develop opinions about whatever it is you're up to. Taking an action can inspire a conversation about why you're making that change. Risk mitigation is all about collective action, so being the inspiration behind a new norm in your community can pay compounding dividends.

can encourage people to create defensible space and harden their homes, it could go a long way toward solving the wildfire problem in communities all over the West.

So far, there aren't any tried-and-true answers, but here are a few that seem to work.

- Cost shares can increase the number of people who mitigate. Studies show that relatively well-off community members are more likely to mitigate based on risk alone. However, people with lower levels of income and education need financial incentives to get the work done. The specifics of what those incentives are may vary widely from community to community, but providing financial assistance is especially important to help the most vulnerable reduce risk.

- Neighborhood comparisons can offer a small degree of change. There are limited data, but one study based on mailers in Colorado showed that comparing one person's property to the neighbors could help target different groups of homeowners who ordinarily might not perform mitigation work on their properties.

- Yard signs can help advertise that a homeowner has done the necessary work. Studies show that when one neighbor has done the work, others tend to do the work too. Wildfire mitigation efforts can spread like, well, wildfire. In this way, community workdays, tool libraries, and other group-centric work can pay large dividends too.

- Other studies show that one factor that can make a difference in the number of homes that reduce risk is simple. If people in a neighborhood talk with one another about wildfire, they're more likely to perform mitigation work. Creating neighborhood wildfire volunteer groups, talking with folks around you about wildfire risk, and organizing community events can all help promote more dialogue.

- There's often a policy window after a fire. The length of that window varies, and the financial, emotional, and human cost of getting there can be tragic. But in the wake of nearby burns, communities are usually more willing to spend resources to address fire risk and work together to prevent future catastrophes.

MORAL HAZARD

In lots of ways, our wildfire problem keeps getting bigger. In addition to our ever-longer fire seasons, the areas most vulnerable to flames—the wildland urban interface, where communities and wildlands meet—keep growing by an area about the size of Yellowstone National Park every year. Today that area is the largest land use type in the U.S. Tens of millions of people live there. Why do we keep building so vigorously where people are most at risk of losing their homes and lives to wildfire?

Answering that question requires diving into economics. From 2010 to 2020, wildfires cost more than $21 billion to suppress. In 2021 alone, that price tag reached a record $4.4 billion for a single year. However, suppression costs are just scratching the surface of the real price tag of wildfires. That's because fires have a whole host of impacts. There are the somewhat obvious, direct costs—things like the loss of private property and the destruction of bridges and roads. But burns also require river and stream cleanup, cause local businesses to lose revenue, and generate effects that are far more difficult to quantify, including the loss of life, long-term physical and mental health effects, and potential loss of a landscape's beauty. The Bureau of Land Management looked at a series of fires across the West in the early 2000s and found that the direct costs were typically less than half of the total cost of a fire and, in some instances, far less. The 2003 Old Grand Prix Pauda Fire in California cost about $61 million to put out, but the total cost of the fire was an estimated $742 million. Wildfire expenses have a long tail.

Who's paying for the enormous cost of wildfires in the West? In the case of the vast majority of major wildfires, the federal government foots the bill. In that way, *you*, the taxpayer, are partly financing the effort. But although the total sum is high, you don't directly feel your contribution; it's distributed among the American public as a whole. A significant part of this subsidy is the vast firefighting apparatus supported by all U.S. taxpayers. This distorts property values because the prices of homes don't reflect the wildfire risk those communities are exposed to. The economists Patrick Baylis and Judson Boomhower studied this effect extensively and estimated that across the U.S. West, the fire suppression subsidy represents an average of 20 percent of a home's value. That's a significant transfer of wealth from taxpayers to WUI homeowners.

In this way, a huge part of the rapid rise of development in the WUI is driven by *moral hazard*. That's a term economists and lawyers use to describe scenarios in which people take on excessive risk because those risk-takers don't bear the cost of their actions. *Moral hazard* was coined in 1862 by a Chicago fire insurance agent, Arthur Ducat. (Yes, the same Arthur Ducat who campaigned for better, fire-resilient building codes in the Windy City.) In an essay over 150 years ago, Ducat suggests that some homeowners with sizable insurance policies might burn down their *own* homes in order to claim an insurance payout. This "moral hazard" exists, according to Ducat, because the policyholder can reap the benefits of a large payout subsidized by other policyholders' premiums.

If you've heard the phrase "too big to fail," you very likely under-stand the idea behind moral hazard. In the years prior to the 2008 financial crisis, banks were making huge bets. When those bets failed, the entire economy fell into chaos, eventually leading to enormous, federal bank bailouts. Wall Street made the bets, but taxpayers assumed the risk.

Kimi Barrett, lead wildfire research and policy analyst at inde-pendent, nonprofit research group Headwaters Economics, thinks deeply about moral hazard and risk in the WUI. The taxpayer bill for suppression is one example of that. But to get a sense of how decision-making across the West plays out, she says it's helpful to appreciate the perspective of, say, a state representative or a county commissioner. "As an elected official, you have no incentive to pre-vent or limit building in wildfire-prone lands when [developing there is] only going to increase your tax revenue base," she says. "If I'm a county commissioner, why would I care if that house burns down? Because I'm not going to pay for it." Growth, no matter the consequences, tends to be the bottom line. The incentives, in essence, are all wrong.

Not all issues that spur development in the WUI are related to moral hazard. Parts of how we got here can be traced back more than 150 years. America has long had a love affair with the West. Starting with the Louisiana Purchase, the U.S. government sub-sidized westward expansion. The government gave land away, land often stolen from the Indigenous people already there, and encouraged settlers to lay claim to their parcel of the frontier. Today, people still seek isolation, solace, and their "claim to the frontier" in the West. Many folks move to the WUI to seek natural beauty—the very feature that endangers them, where homes and wildlands meet. Those areas also offer other amenities, including the space required to eke out a family's version of the American dream. Recently, COVID-19 enabled a boom in remote work, and some people have taken advantage of that luxury to leave the city for scenic locales.

At the same time, people also tend to underestimate their own risk. A Forest Service survey in Jackson, Wyoming, one of the wealthiest WUI towns in the U.S., found that only 10 percent of homeowners rated their wildfire risk as high or extreme. Wildfire experts, however, looked at the homes of survey respondents and rated 46 percent of them as high to extreme risk. From a home-owner's perspective, it's easy to ignore the reality of wildfire risk—to think "Oh, that'll never happen here." But the reality of risk becomes stark when a fire hits a community.

Other folks move to the WUI because it's their only option. City centers are becoming increasingly expensive. People look to other areas, in the mountains and foothills, as more affordable alternatives. Instead of being drawn in by the area's natural won-ders, they're often forced there, with nowhere else to turn.

Altogether this creates a cycle of more development in the WUI that will be hard to reverse. "There are no examples of us as a society saying you cannot develop there because of all the fire risk," Barrett says. "We're thinking about that with sea level rise and with flooding…we're not really doing that with wildfires." Barrett sees two possible paths that could reshape how and where we build.

One option is proactive: citizens and policymakers recognize that there needs to be a change. That means acting before more disasters occur. The dangerous cycle of development in the WUI occurs at the intersection of climate change, affordable housing, and wildfire. Addressing this fire-prone area could be an oppor-tunity to take policy actions that address all three issues. It could mean placing more strict limits on where development—and, after fires, *re*development—can occur. It could also mean helping families with "managed retreat," or providing federal aid to move homes to safety, away from areas vulnerable to natural hazards like wildfires and floods.

The alternative is a reactive approach. Intense fires will keep burning communities, Barrett says. In a somewhat morbid sense,

TOOL LIBRARIES

Libraries are the ultimate community resource. They are one of the few institutions almost all Americans value. A Pew Research study found that over 90 percent of Americans think libraries improve our communities and play an important role in giving everyone a chance at success.

As communities in the West come to grips with mitigating wildfire risk and, in an increasing number of cases, rebuilding, libraries are emerging as a critical resource. But these libraries don't lend books, they lend tools. If you've ever attempted a home DIY project, you likely know that having the right tool for the job is critical. And often that right tool is expensive—difficult to justify for a one-time use. And tools can take a lifetime to collect.

In places like Santiam, Oregon, and Boulder Creek, California, community-minded folks have started tool libraries. They are usually simple—a shed or container that offers a place for people to donate tools to share. Businesses pitch in. People can borrow those tools free of charge. Those include things like chainsaws, rototillers, post-hole diggers, rakes, and other specialized tools you might need to clean up and move forward.

Resources like these are more than practical. They build community and demonstrate that we're in it together.

she and many others in the wildfire community are waiting for a transformational moment—a devastating event that will show just how high the stakes are to continue business as usual. Lots of people thought that the Camp Fire would do exactly that, when eighty-five people died. But little changed. "So what will it be?" Barrett says. "That's the big question. It's going to have to be bigger than the Camp Fire, which is pretty terrifying to think about."

ZONING, REGULATIONS, AND THE WILD WEST

While creating defensible space, replacing roofs, and the like create the basics of individual homes' fire adaptations, instituting rules and regulations to ensure new and existing houses are up to snuff when it comes to wildfire can be key to community-wide resilience. That's much easier to achieve in theory than in practice, though.

For communities and community members interested in changing what's happening locally, here's a short primer on best practices in wildfire zoning and regulation.

- **WUI overlay zones**
  This basically means zoning requirements specific to the areas most at risk of wildfire in a given city or county. These zones can require or encourage fire-resistant building materials and specific landscaping requirements—like establishing defensible space around homes and structures.

- **Subdivision regulations**
  These can make sure new developments are as fire-safe as possible through mechanisms like ensuring subdivisions have multiple access points, homes are set back an adequate distance from potential hazards, and vegetation stays under control. They can also mandate open space within new development that could buffer the area from fire.

- **International Wildland-Urban Interface Code**
  This code was developed in 2003 by a global group dedicated
  to building safety called the International Code Council. The
  ICC updates this particular code every three years, though local
  areas often introduce amendments of their own when adopting.
  In addition to rules about driveway width, access points, and
  water supply, the 2021 IWUIC dictates that new homes must be
  built with ignition-resistant materials.

- **Vegetation management plan**
  This is part of the IWUIC, but warrants its own explanation.
  These are essentially detailed plans of attack to reduce fuel on
  newly developed (or, at times, existing) sites, and to maintain a
  fire-safe landscape around the house.

- **Open space planning**
  These measures have often been put in place for aesthetics and
  recreation—but preserving swaths of open land can also help
  create fuel breaks between communities and wildfires (assum-
  ing, of course, those lands are maintained to make sure they
  don't get overgrown). New open space often comes from conservation easements on private property, sometimes purchased
  through bond measures by the city or county. Those bonds
  would generally have to be approved by voters.

- **Steep slope ordinance**
  These rules can prohibit development on hillsides where, due
  to their steepness, fire is likely to rip through if it enters the
  area. The orientation of buildings on these features also often
  makes exit difficult during an evacuation and access difficult for
  authorities.

- **Density bonus**
  These measures allow more homes per acre on a given
  development, if certain conditions are met. In particularly

wildfire-prone areas, this could be a dangerous mechanism. But localities can disallow density bonuses in the WUI and use them strategically, to encourage more fire-safe development in areas that are resilient to hazards like fire. These are also often a tool to promote more affordable housing in areas in need of it.

This list isn't exhaustive; it's just a starting point. One downside here is that many of these mechanisms target only new development—or property owners interested in adding onto or substantially changing their current residence. That leaves out a huge part of the population: namely, people already living in areas vulnerable to fire in need of some major retrofits. Communities across the West are experimenting and figuring out how to get their constituencies on board with these kinds of rules.

San Diego, California, for example, has an enormous WUI—almost five hundred miles long, with tens of thousands of people residing there. The city has a rigorous brush management policy to thin and control weeds within one hundred feet of structures. If homeowners don't do the work, they have seventy days to comply before the city hires a contractor and sends the property owners the bill. Eagle County, Colorado, too, requires vegetation management plans for any new and special use development. New construction anywhere with fire danger of moderate or higher is also required to use specific fire- and ember-resistant materials for roofing, siding, decks, and soffits. In Ada County, Idaho—home to Boise—new development is required to have fifty-foot defensible space around homes. Teton County, Wyoming, and Flagstaff, Arizona, have both adopted the IWUIC, albeit with amendments that fit the needs of their local areas.

This is just a smattering of localities attempting to regulate wildfire risk mitigation across the West. It's important to note that city planners rarely rank as the most beloved of government employees in any community. Generally, getting a permit to do what you want to do is onerous and expensive, and sometimes impossible. City planners are often seen as gatekeepers, standing between a homeowner and some desired outcome. At the same time, many of the areas that have adopted proactive wildfire regulations are relatively liberal, urban, and wealthy. That's likely in part because there's fierce resistance to regulations in much of the rural West. In these areas, "Don't Tread on Me" flags fly, and a general ethos dominates: you mind your business, I'll mind mine. This libertarian streak runs deep in western politics, and with it comes an aversion to rules and regulations and a skepticism of government. The ethic often focuses on individual, rather than collective, responsibility.

In terms of wildfire preparedness, take a look at Oregon. In 2022, the state mapped the communities in the state that faced the most vulnerability to wildfire. A bill was passed that would help fund home mitigation work in the areas that most need it, and the legislation required that mapping effort. When the map came out, the state mailed notices to every homeowner in high-risk zones. That's when the backlash began. In the weeks that followed, the state received thousands of angry public comments. More than fifteen hundred homeowners appealed their risk rating. While the step seemed relatively benign, people worried the step represented a slippery slope to government overreach. For the state, it was a fiasco. Oregon gave in and withdrew the map. This example shows that science and maps, especially when coupled with politics, aren't enough. The wildfire problem is also a people problem.

However, there is room for optimism in getting these sorts of regulations on the books. Pew Research polling in 2021 shows that, nationally, 62 percent of U.S. adults—a sizable majority— believe that the government won't go *far enough* in limiting new construction in areas at high risk from climate hazards like storms, floods, and wildfires. Only 33 percent were concerned that the government would go too far. The results were somewhat reflective of party lines; far more Republicans than Democrats, for example, worried about federal overreach. But the results also show that there's broad public demand for proactive regulations. The same poll, in fact, showed that about half of adults in the country ranked setting stricter building codes to reduce the damage of climate catastrophes as a "very important" goal to them. Another 37 percent said it was a "somewhat important" goal.

Bridging that gap—between public safety and the desires of individuals for freedom and lives free of interference—is the crux of getting these regulations on the books. One solution involves incentivizing the work. Some areas experiment with cost-sharing programs to help property owners—voluntarily—retrofit their homes with fire-resistant materials and establish defensible space. Even bigger picture, the key is community building. Educational campaigns and events are a start. But cities and counties need to show genuine interest in the people who live there. Community members need to feel like their concerns are being heard, that local officials have their best interest in mind, and that regulations are meant to *help* rather than hinder.

There's no one-size-fits-all solution for every community in the West. And many areas will likely never find the political will to enact rules and regulations. But that doesn't mean there's no hope. As more bright spots emerge across the West, more organizations will become dedicated to finding planning solutions in the wildfire space. One example is Community Planning Assistance for Wildfire, an interdisciplinary set of resources for local land managers, fire departments, and land use planners. It was launched in 2014 by Headwaters Economics and Wildfire Planning International and has funding from the Forest Service. This group has expertise in fire science, land use planning, public policy, and forestry and provides free advice to communities in need. If you live anywhere near wildfire risk, your local officials should know about CPAW. If they don't, make sure you tell them about it.

WHO SHOULDERS THE RISK?

About fifty million homes are in areas vulnerable to wildfire today—a number increasing by one million every three years or so. At the same time, wildfires are burning more land and affecting more people than in living memory. So who's most at risk?

A number of studies have looked at who lives in the country's most wildfire-prone areas. Spoiler alert: white people make up far and away the largest proportion of wildland urban interface residents. And that makes sense. The typical image of the WUI—or the epicenter of the wildfire problem—features fancy houses, often second homes for wealthy folks seeking the natural wonders of the world. It follows that most people conceive of wildfire as a problem affecting mostly well-off white folks. But a deeper look at who actually lives in the WUI and how wildfires affect them reveals a much more complicated story, one that illuminates some all-too-familiar problems of inequity in our society.

A 2021 study, for example, used census data to examine areas in California burned by wildfire from 2000 to 2020. Over those twenty years, the authors noted that the number of people affected by fires almost doubled, while the overall area affected went up by about twenty-three thousand acres per year. Critically, rural areas were affected much more severely than their urban counterparts. In those rural areas, the data reveal, rates of poverty and unemployment were relatively high, and people tended to have relatively low income and fewer college degrees. People over the age of sixty-five were also much more at risk. That's particularly significant because the elderly often have a harder time evacuating and are more likely to have preexisting health conditions, making them more vulnerable to wildfire smoke. Their findings, the authors said, suggest that wildfire is an environmental justice issue—that is, the effects of fire aren't borne equally. "Disadvantaged families who cannot afford to live in urban areas are rendered at greater risk of dangerous wildfires that may impact their health and further exacerbate socioeconomic inequities," the authors wrote.

Another study, released in 2018, tracked how capable of withstanding wildfire communities throughout the U.S. were by looking at a collection of thirteen traits, including things like English fluency, health, and income. The authors layered that data on top of wildfire risk maps. Their results showed that communities that were majority Black, Hispanic, or Native American were about 50 percent more vulnerable to wildfires than predominantly white communities. Native Americans, in particular, were about six times more likely to live in areas susceptible to high-intensity fire. The upshot means that when wildfires occur, their impacts are often borne along existing social fault lines. That means that society's most vulnerable might feel the worst effects and take longer to recover.

Zooming in can help make sense of these results. Relatively wealthy folks who choose to live in fire-prone areas can often afford the mitigation measures necessary to lessen risk. They have insurance, jobs, and, importantly, free time. Outreach measures often target these populations too. A research team at a nonprofit called Resources for the Future looked at the distribution of wildfire risk reduction services after fires—important actions like thinning forests and reducing accumulated debris and other flammable stuff. In the aftermath of a burn, the whiter, higher-income neighborhoods are the most likely to get the Forest Service's attention. Communities with more resources can more effectively lobby for assistance. These people are more likely to call their elected representatives. Thinning projects and prescribed burns are expensive to execute, and federal and state governments are unable to treat all the areas that need them. Given this scarcity, the loudest voices get the most help, and those voices tend to be relatively wealthy and white.

The impacts of certain fires also tell disturbing tales about what burns can mean for equity. Fires in 2020 in Oregon and Washington, for example, left thousands of migrant workers without homes. In 2014, language barriers during another Washington

FIRES AND FLOODS

The effects of fire don't end when the flames go out. Some of the most devastating impacts of fires link back to erosion. On hillsides, the trees, shrubs, and especially their roots act a lot like a net, holding the soil in place. When high-intensity fire burns through, it can sear those ties that secure the soil. The biggest, hottest of fires can also burn the soil itself, making it *hydrophobic*. That means that rainfall will sheet right off the ground, almost as if it were pavement.

When rain comes, it can be devastating. All that sediment can wash into municipal drinking water. Stabilizing soil and dredging water storage from sediment can cost tens of millions of dollars. Postfire erosion can also lead to landslides and floods. This became immediately apparent in early 2018, after the Thomas Fire had burnt nearly three hundred thousand acres near Santa Barbara, California, the year before. The new year brought with it four inches of rain over two days. On the night of January 9, a twenty-foot-tall wall of earth descended on the small hamlet of Montecito, destroying or damaging over four hundred homes and claiming at least twenty-three lives.

While this example was particularly tragic, postfire landslides have become commonplace in parts of New Mexico and Arizona. FEMA recommends purchasing flood insurance if a fire has burned nearby recently. Some erosion control interventions, like using logs or straw mats to stabilize soil, can also help mitigate the chances of a devastating landslide.

fire prevented predominantly Hispanic farmworkers from receiving evacuation notices. The only Spanish-language radio station nearby never received those notifications. During California's 2017 Thomas Fire, a similar story unfolded: local radio stations weren't able to release timely, accurate bilingual information. After the burn itself, studies showed that officials ignored the needs of the very same immigrant communities that were neglected as the fire approached. In Sonoma County in 2017, price gouging after a wildfire made an existing housing crisis much worse.

All impacts of wildfire might be harder to cope with for low-income community members, who are less likely to be insured, have fewer resources available to rebuild or relocate, or are more likely to be renters and therefore ineligible for many federal and state rebuilding programs. Those issues compound for members of minority groups, who, due to historical legacies of abuse and mistrust, also might be suspicious of state and federal organizations offering help.

And it's not just the flames themselves. The effects of wildfire smoke present unequal effects as well. Research from the University of Washington found that disadvantaged communities are at highest risk when it comes to pollution, including wildfire smoke. These folks often work outdoors and live in older homes more likely to let wildfire smoke creep in. They are less likely to have air filters and other mechanisms to clean the air in their homes. Taken together, these communities are therefore more susceptible to all the bad downstream health outcomes that long-term and intense exposure to wildfire smoke can bring.

Unwinding these inequalities will be hard. But it's also an opportunity to recognize that a multitude of crises affecting the country aren't occurring in isolation. Addressing the impacts of wildfires requires addressing climate change, for example. And efforts to address both of those are for naught without meaningful efforts focusing on race, justice, and the people most often left behind.

UNINSURABLE

Nobody likes paying for insurance—until they need it.

Across the country, insurance companies are scrambling to figure out how to handle wildfires while homeowners bear the consequences. In some cases, companies are flat out refusing to insure homes. California, in particular, is testing the limits of the insurance industry. It's worth understanding what's happening there to make sense of what might unfold across the rest of the country.

About three times as many homes in California are at risk of wildfire as in any other state. At the same time, the state has seen a series of devastating fire seasons. In 2017, more than 1.5 million acres burnt. The extent of the devastation was record-breaking at the time, and forty-seven people died in the burns. The next year, more than eight thousand fires burnt, just shy of 2 million acres in the state. Those burns included the Camp Fire, the deadliest in the state's history. For many, it seemed like fires couldn't get worse. But 2020 was another year for the record books. More than 4 million acres burnt, destroying about ten thousand buildings.

This presented a real dilemma for insurance companies. By design, they rake in cash when people *don't* need their insurance. When homes are relatively safe from serious damage, insurers thrive. And for decades, this was more or less the case for homes across the country. There was no persistent threat from fire built into the system. Then, things began to change—fast. From 1964 to 1990, U.S. insurers paid out on average about $100 million a year for wildfires. From 1990 to 2010, that number rose dramatically: to about $600 million a year. But the worst was yet to come. Over the next eight years, that number exploded: insurance payouts averaged nearly $4 billion annually. As the climate got warmer and drier, as a century of fire suppression loaded the country's forests with fuel, and as more and more people moved to on fire-prone lands, the risk became built into the system.

The epicenter of the insurance companies' conundrum was California. Part of the problem was that, under state law, insurers could take into account only past losses in their rates. There was no way to factor climate emergencies, likely to get even worse in the years to come, into their pricing analytics. Insurance companies responded the only way that made financial sense at the time. In certain areas, they simply refused to renew insurance policies— or to offer insurance at all. The ten most fire-prone counties in California experienced a 203 percent increase in nonrenewals from 2018 to 2019.

For homeowners, the impact was sudden. Homes store precious memories and experiences; they're extensions of the people who live there. Very often, homes are their owners' single most valuable asset, and losing a house would mean financial and emotional ruin. Lives and livelihoods were at stake. Unable to get their insurance renewed or to find insurance to begin with, many Californians had to turn to the state's Fair Access to Insurance Requirements, or FAIR, plan. Similar plans are on the books in thirty-one other states; it's meant to serve as a last resort for homeowners who can't get insurance anywhere else. However, the measure is meant only as a temporary stopgap. It's much more expensive than traditional home insurance options and covers only basic losses.

In 2019, the state stepped in. The insurance commissioner issued a one-year moratorium on nonrenewals in wildfire areas statewide. Additional yearlong nonrenewal moratoriums for particular wildfire-adjacent areas followed in 2021 and 2022. They were piecemeal actions taken by state government to alleviate the pain of homeowners in disaster zones. They paid off, at least to an extent. In 2020, for example, a California Department of Insurance report found that statewide nonrenewals dropped by 10 percent in 2020, compared with the year before. Eighty percent of that decrease was due to the moratorium. However, as nonrenewals were going down, enrollment in the state's FAIR plan was going up. In 2020, about fifty thousand additional Californians turned to the program, having no other options for insurance. Its enrollment reached record highs.

Across the entire state, homeowners' insurance rates increased an estimated 10 percent from 2021 to 2022. Some insurers bailed on California altogether. As of 2022, Geico had closed all its physical offices in California. AIG and Chubb also cut back dramatically on the number of California homes they would insure, with AIG dropping a reported nine thousand homes. Those insurers that stay might find less overt paths to stay profitable. One 2022 watchdog organization report alleged that insurance companies were finding legal loopholes to gouge customers—like refusing to compensate homeowners for smoke damage and giving homeowners unrealistic timelines for filing claims, among many other violations of California law. Surveys in the wake of wildfires like the Camp Fire have consistently found that the majority of homeowners affected by the flames were—usually unknowingly—massively underinsured.

In hopes of changing the playing field, California's Department of Insurance issued a potential game-changer. A new rule in 2022 requires insurers to grant premium discounts to policyholders who make fire mitigation and safety investments at their property. Basically, if homeowners do the work necessary to make their property more resilient to fire—like clearing flammable material around homes and installing fire-resistant materials such as metal roofing and cement siding—they won't have to pay as much for insurance. It's the same idea that Wildfire Partners took hold of in Boulder, Colorado. This gives homeowners incentives to make changes and helps insurance companies rest a little easier, knowing that their customers are taking proactive steps to mitigate fire risk. There are lots of qualifiers that come with the new rules, and lots of unanswered questions. One particularly crucial question to answer will be: What resources are available to those most in need, who can't afford or don't have time to make changes on their properties?

The California insurance market will almost certainly continue to evolve into the future. What happens there is likely to spread to other states in the West.

A CREATIVE SOLUTION

As the case of California demonstrates, the insurance industry is facing a crisis. And that's a crisis that affects millions of people across the country. Insuring against disaster in an era of climate change requires creative—and proactive—thinking. Here's one solution that has potential to revolutionize insurance for wildfire, along with a myriad of other climate hazards.

It's somewhat intuitive that the people who shoulder the worst of the risk ought to pay more for their insurance. That's how the National Flood Insurance Program operates, for example. Folks who live in a floodplain are required to purchase separate flood insurance, often at great cost. Certain areas at especially high risk of flooding are prohibited from being developed entirely. The program has its pitfalls, yet it could follow that a similar fire-related insurance program, run by the government, would be one way to move forward.

But looking at what climate change means for all kinds of natural hazards—from floods to wildfires to hurricanes—some scholars are thinking even bigger. Their grand idea is to pool all those hazards together. Then, everyone would have to buy in, distributing the risk more evenly across the country and helping buffer insurance companies from the blows of constant payouts. This form of disaster insurance could also place limits on where we build and how we build. Premium prices and deductible levels could be used to incentivize flood-, fire-, or earthquake-resilient construction, stricter zoning, and other risk-mitigating behaviors and investments like clearing brush or installing flame-resistant roofs.

A program like this would undoubtedly suffer immediate political blowback. But this is one potential solution, which recognizes that no one is safe from natural hazards and that, in an era of climate change, we're all in this together.

RISK MODELING

How do you know how much risk your property might face? The U.S. Forest Service maintains a website, Wildfire Risk to Communities (wildfirerisk.org), which has comprehensive risk mapping covering the entire country. The target audience, however, is policymakers and land managers. Risk is represented at the community level. The website is certainly interesting and illuminating, but not all that useful for an individual homeowner. In 2022, the First Street Foundation, a climate and technology non-profit organization, released a Risk Factor calculator for wildfire risk. At their site, riskfactor.com, you can search any address in the US and get a risk factor score and an explanation of how it's derived. In addition to the current risk score, Risk Factor also estimates how wildfire risk will change over the next thirty years on a variety of dimensions. This sort of individual-level data is a great way to understand what risks you can mitigate and what risks you cannot. Note, however, that these kinds of tools are difficult to maintain. Conditions on the ground change as fires alter the landscape, folks make changes—for better or worse—to their homes, agencies conduct fuel treatments, and neighbors do things that increase or reduce risk. So this is just a start.

WILDFIRE TAX

Replacing all the roofs in the WUI nationally in need of fire-resistant material would cost at least $6 billion. All the fencing, decks, vents, and more—plus the millions of hours of labor required to reach the scale we need—would add billions more to that price tag. Who should pay? Should it be homeowners themselves, the federal and state governments, or some other financial scheme?

Places all over the West are trying to figure that out. In 2022, the state of California had a massive but potentially game-changing measure on the ballot. Known as Proposition 30, the initiative would have put an additional 1.75 percent income tax on anyone making over $2 million in the state. Basically, the idea was: tax the 1 percent. Proponents said that it could generate up to $5 billion annually and would go toward climate resilience in California—especially things like EV infrastructure. About 20 percent of the funds generated would pay for wildfire prevention and suppression.

When November came, more than 57 percent of voters checked "no." The proposition failed partly because it divided Democrats; California's governor came out against the measure, and some environmental groups, like the Sierra Club, opposed it because it could promote logging. In the end, it was an innovative approach that might have funded millions for all things wildfire: from suppressing and fighting fires to preventing them before they occur to making homes more resilient. Earlier in 2022, Napa County residents specifically had also voted no on a quarter-cent sales tax that would have gone toward wildfire prevention. The area had experienced devastating fires in two of the last five years. These two cases show that after years of devastating fires, California has an appetite for finding new mechanisms to deal with its wildfire problem. The hard part, however, is finding the political will and public acceptance to do so.

Boulder County, Colorado, tried a similar approach. In 2022, voters there also had a wildfire-related measure to decide on. Ballot

Issue 1A proposed a 0.1 percent sales tax in the county. That would generate an estimated $11 million in its first year, entirely going toward wildfire risk mitigation—work like thinning and prescribed burning in forests, as well as replacing roofs, establishing defensible space, and implementing the other efforts required to harden homes.

The measure passed overwhelmingly. More than 72 percent of voters said yes. The burden of the tax was relatively small—only a penny on every ten dollars—and it was evenly distributed. Less than a year earlier, the Marshall Fire had raged in the suburbs of Denver. It was the most destructive in the state's history. With the evidence of what wildfires could cost communities so fresh in mind, voters likely sensed the urgency of the problem.

Many counties, cities, and states across the West are looking to this experiment on the Front Range of Colorado. Wildfire in the WUI is a vexing problem at an enormous scale. The success of Boulder's sales tax might inform how other areas in the West approach fire mitigation work in their own backyards. These questions are hard and fraught with all sorts of ethical and political dimensions that vary from place to place. Still, much as we need to find a better way to live with fire, we need to find a better way to share the costs.

ADAPTING IN ACTION

The Durango & Silverton Narrow Gauge Railroad has run continuously since 1881. It originally moved gold and silver from mines in southwestern Colorado's San Juan Mountains. Today, it operates as a heritage railroad, moving tourists between these two mountain towns. Historically, trains were a common cause of wildfire; it was normal practice for water carts to follow steam engines down the tracks, dousing fire starts along the way. And some things never change. On June 1, 2018, an ember from a train on that railroad sparked what would become the "416 Fire." It burnt for over sixty days in the San Juan National Forest, torching more than fifty-four thousand acres. Fire had been suppressed in the area for more than one hundred years, and the flames burnt with ferocity. They were heading straight for Durango, with a population of about twenty thousand. However, the fire was contained and never reached the town—at a cost of more than $25 million—and no homes were lost.

A key reason has a lot to do with how one tiny community responded to a fire sixteen years prior. In 2002, the Missionary Ridge and Valley Fires combined tore through more than seventy thousand acres. They scorched more than forty-five homes. It was a fast-moving, unpredictable fire. Flame heights were reported at over 250 feet in the air. A few of those homes were in Falls Creek Ranch, a community of about forty-five people today, which sits on around nine hundred acres just northwest of Durango. When residents returned home after the fire, they saw brightly colored flagging on their properties. Red meant firefighters were unable to access the area and safely do anything to protect the property; green meant they were. After the fire, Durango fire chief Dan Noonan told the public, "If your home was unsafe for firefighters, we couldn't risk lives to protect it."

Those bright, almost neon, reminders made the scale of the problem immediately tangible to property owners. It was a wake-up call; the work began almost immediately. Soon, leaders in the homeowners association sought guidance from Firewise USA, the National Fire Protection Association's nonprofit program that helps individuals and communities develop wildfire action plans. Residents began clearing brush and hardening their homes. Community cleanup days were organized—three per year—during which residents worked together to gather pine needles and clear underbrush. They started holding regular educational events and practicing evacuation drills. Wildfire risk mitigation and collective action became embedded in the norms of the community. Before long, they'd amassed an estimated thirty-five hundred annual service hours.

In partnership with a slew of state and nonprofit agencies, Falls Creek Ranch signed a Community Wildfire Protection Plan in 2011. This fifty-four-page document, chock-full of acronyms, might seem like a bunch of government red tape. However, establishing a CWPP is a big deal. It puts communities on an expedited path for federal fuel reduction operations.

All this organization and work paid off in 2018. When the 416 Fire blew up, Durango Fire and Rescue chief Hal Doughty realized that Falls Creek represented more than just a bunch of homes that were more resilient to flames. It was an area perfectly positioned to help firefighters get a handle on the flames. It was an asset rather than a liability. Fire crews could save the neighborhood—but that was just the start. They could leverage its fire mitigation and fuel reduction efforts to stop the 416 fire's march toward Durango. Beyond a safe, relatively fuel-less zone in which to operate, Falls Creek Ranch gave firefighters another critical resource: time. It gave them the time to allocate resources to other areas and priorities. In a dynamic, rapidly advancing wildfire scenario, time is an invaluable asset.

All wildfires are complex and influenced by a multitude of factors. By nearly all accounts, the community effort in Falls Creek Ranch paid massive dividends. No homes were lost, and the 416 Fire stopped short of Durango. Many aspects of wildfire compound to exacerbate problems. Climate change, development in the WUI, and long-term fire suppression conspire to create more dangerous fire conditions. But the actions we take to mitigate wildfire risk compound as well. The whole of a neighborhood that mitigates its risk is greater than the sum of its individual homeowner efforts. Through collective action taken to protect their own neighborhood, Falls Creek Ranch residents gave firefighters an opportunity to protect the much larger town of Durango. When the dust settled, Chief Doughty explained that Falls Creek Ranch was a pivotal neighborhood. What happened there, he said, was "the fight that won the war."

TURNING OUT THE LIGHTS

For decades, utility providers used so-called rolling blackouts to manage times when energy use was surging, putting too much of a demand on the electric grid. In short, when too many people are using too much power, providers turn out the lights. These often occur during periods of extreme cold or heat and can be at best annoying and at worst deadly. But wildfire is presenting a new challenge for the country's energy system.

Power lines present significant wildfire risk. From 1992 to 2020, the National Interagency Fire Center reported more than thirty-two thousand wildfires started from power lines across the country. Some of those have been the largest and most devastating fires in living memory. The nearly one-million-acre Dixie Fire started from a downed Pacific Gas & Electric power line. So did the Camp Fire, the deadliest in California's history. In that case, PG&E pled guilty to eighty-four counts of involuntary manslaughter after the burn. Flammable stuff like tree branches tend to grow into contact with lines, and the lines themselves need constant maintenance. The towers tend to be the highest things around, attracting lightning strikes. High winds can knock them over. Research shows that fires caused by power lines tend to be larger and more devastating, most likely because they ignite closer to communities. Given the risk, power companies throughout the West now routinely shut down their power systems during times of extreme wildfire risk, a practice they call Public Safety Power Shutoffs.

Those shutoffs effectively get rid of the fire risk from power lines. Live wires go dead. But this strategy necessitates severe trade-offs. Shutting down the power to millions of people is far more than an inconvenience; it can be dangerous. Many people need to power medical devices in their homes to survive. Lots of folks rely on air conditioning to endure the sort of intense heat that frequently accompanies extreme wildfire risk. Think of the public safety that comes with functioning streetlights. Shutting

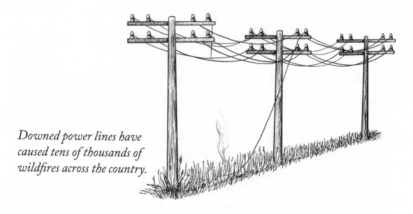

Downed power lines have caused tens of thousands of wildfires across the country.

down power might force evacuations, another communications and resource-intensive operation. Beyond these potential costs, the challenge of informing huge numbers of people of an impending power outage is immense. And yet, if a catastrophe like the Camp Fire can be avoided, are blackouts the least bad option?

These shutoffs aren't going away any time soon. It's best to prepare ahead of time for when they occur. Here are some tips to keep you and your family safe when the lights go out:

- When wildfire risk is extreme, especially in California or Oregon, get ready for a blackout. If you have medical devices or special needs, contact your power company and register a Medical Certificate. That helps make sure they contact you when a shutoff is coming. If relocating is difficult, consider a backup generator and make sure it's set up for safe operation. Generators are often fueled with gasoline or diesel, so be careful where you put them and when you run them.

- Keep your phones and other electronic devices charged. Get some backup chargers and stock up on batteries for flashlights, headlamps, or anything else you might need with the power off.

- Top off your car's gas.

- Make some extra ice and store it in your freezer. When you know the shutoff is coming, stock a cooler or two with ice and the food you plan to eat in the immediate future. Leave other food in the fridge or freezer and don't open either if you don't have to.

- Check on your neighbors, especially the vulnerable ones, and help them get prepared. Make sure the power company has your contact information and stay attuned to any updates.

- Whenever possible, keep the mood light. If you have kids, power outages can be fun—it's camping without the hassle of lugging a backpack or swatting mosquitos.

INNOVATION FROM ADJACENT SPACES

Sometimes it's necessary to look outside a particular domain for fresh ideas. Hollywood makeup artists inspired important advances in surgical safety. Glass artists developed techniques that enabled the development of eyeglasses and microscopes. Scholars of technology call this "innovation from adjacent spaces," and in at least one case, it could upend how people think about wildfire in communities.

Hussam Mahmoud, a professor at Colorado State University, studies civil engineering, with a focus on bridge strength and how buildings handle earthquakes. When his research team landed a grant for looking at fire resilience, he didn't know much about fires, but he knew lots about structures. He started to wonder why a structure would or would not catch fire. He was dissatisfied with the existing maps of wildfire risk. Many of them just look at risk in general—as in the risk level of a single parcel of land or community. Or else they stick to the forest, looking at fire spread under different scenarios based on fire behavior, fuel, and weather. Those models often produce the probability that a fire might *reach* a community.

However, the analysis more or less stops at the outskirts of neighborhoods, with the end of the trees and shrubs. Imagine embers drifting from an intense fire a mile away, igniting the roof of a single house. What happens next? There was this major piece missing—mapping how fires spread once homes themselves become the fuel. "Once the fire gets inside the community, we have no idea how much damage there will be," Mahmoud says.

To fill this gap, Mahmoud decided that more detailed data were needed. That meant looking at likely wind behavior, where homes were oriented, how they were laid out and distributed within the community, and what the surrounding fuel was like. To make sense of all these factors, the research team borrowed a mathematical theory often used in public health research on the spread of infectious diseases. This particular innovation from an adjacent space—disease networks—enabled Mahmoud and his team to look at certain homes as "superspreaders," or structures that, once burning, might be more likely to ignite other homes nearby. Once that first home starts to burn, Mahmoud says, it's not just new embers from the fire that might cause other homes nearby to go up in flames. Radiant heat also plays a role; the sheer warmth of new burns might ignite nearby structures.

As a whole, the superspreader analogy holds up well for fire spread in a community, Mahmoud suggests. The research team was more or less able to map the community's "immune system," he says. The spread of fire from home to home is somewhat equivalent to contact tracing. Creating defensible space? That's a bit like social distancing. Replacing roofs and vents—all the work required to harden homes—functions a bit like a vaccination. "I think viewing the wildfire problem in this context is actually important to start to think of what kinds of models we really need," Mahmoud says.

The team has honed the model over years, comparing it with the impacts of real-life burns like the 2018 Camp Fire and the 2020 Glass Fire that ignited in Napa County. It's tested as high as 86 percent accurate at predicting which homes will burn and which homes will not.

Mahmoud and his team's model is especially important in a world of scarce resources. It can identify the buildings or neighborhoods that are especially important to harden. Those "super-spreaders" might hold risk for what else would burn should they ignite. The modeling effort recognizes that wildfire mitigation work isn't one-size-fits-all, and it could give communities a head start in prioritizing their efforts.

AFTER THE BURN

LOSING A HOME

When a fire burns through a community, it can be absolutely devastating. Some call it "community-wide trauma."

Rod Moraga experienced what that's like firsthand. Moraga has been a firefighter for decades. He started when he was in college in New Jersey. He took a forestry internship in Colorado one summer, and he remembers spacing out during the wildland fire part of his orientation. At the time, he had no intention of fighting fire. But that was 1988. It was the year more than a million acres burnt in and around Yellowstone National Park. Those conflagrations were a national emergency; it was all hands on deck. Moraga was sent to contain the flames.

Those first weeks on the ground, digging fireline, he absolutely hated it. "The reason I go to college is so I don't have to dig a ditch," he says he thought at the time. "That was the whole reason I thought I was getting an education, to avoid hard labor." But in that misery, he found a little bit of solace in the flames themselves. He was struck by the power of flames as he saw them burning through the landscape. He was fascinated. Despite his initial reluctance, he's stayed in wildland fire ever since.

Fast-forward to the early 2000s, and Moraga and his wife were living in Boulder, Colorado. He says the cost of living was sky-high. They wanted to purchase a home, but to make it happen, the mantra was "drive until you can afford." In their case, that meant heading up into the foothills of the Rocky Mountains. They found a spot, and they fell in love with it. The roads were steep and rutted. It was isolated, but there was also a sense of community. That was part of the appeal. Nobody locked their doors. If he ever needed a tool, he could go knocking next door. There was no cell service. At night, he loved to soak in his hot tub and stare up at the stars. As a fire-fighter, Moraga was also acutely aware of the fire risk, and he'd done some mitigation work on his home, like installing a metal roof.

In 2010, he was part of the local volunteer fire department, the first responders to any flames that broke out nearby. He got a notification that there had been a fire start in the area. When he arrived, it was tiny: less than half an acre on a hillside. But then, he says, the winds came. The fire exploded. By day's end it was burning more than five thousand acres.

Moraga was in charge of deploying resources—things like fire engines to help contain the flames. Meanwhile, his wife, his son, his son's friend, their two dogs, and their two cats piled into their car to head out of the canyon. It was clear that the danger was getting near. With his family safe, Moraga felt like he needed to check on his house when he had time. He drove up the road, and all seemed fine. He kept on attending to the rest of the fire.

But before long, he heard that the fire had "spotted"—meaning flying embers had created a new swath of flames. And that newly burning area was right by his house. He went back up there, and this time, he says, "there was just ash and embers falling everywhere." He saw some smoke at a neighbor's house—a piece of the deck was on fire. He flagged down an engine to put out the flames. It saved the home, and so the engine left to head to a different part of the fire.

Eventually, he came back to his house for a third time. The scene had changed, Moraga says, and low-intensity flames were creeping down a hill toward his house. He says that it was a frustrating moment. Had there been a fire engine there, he could have likely put it out. But Moraga was in charge of operations and had deployed all those resources elsewhere. "A little bit of irony," he says. Helplessly, he watched. Soon, his deck was smoking. "That's when I knew it was gone," he says. He watched the fire start to swallow his house. When it was about half in flames, he left. He says that he had to try to do what he could for the rest of the fire.

In the end, the Fourmile Canyon Fire burnt more than six thousand acres and nearly 170 homes, including Moraga's. More than three thousand people were evacuated. Eventually, Moraga drove back up those rutted dirt roads to look at the remains of his house. He still has the pictures. His metal roof sits, intact, on top of the ash and debris.

Moraga says that he had emotionally hardened over the years. It was a requirement of the job. He'd witnessed many homes lost to fire. Of course, he says, people confronting the flames always feel terrible when homes are lost. But it's important to focus on what you can do, not on what's already over and done with. "As firefighters," he says, "we just can't get all emotional in the midst of our job or we wouldn't be able to do it." Suddenly, on the other side of it all, he had a very different connection to what it meant to lose a home to flames. "There's a lot more than four walls involved in a home," he says. "It can be difficult at times."

At the end of the day, Moraga felt lucky. He lost his home, but his family and his pets—the things that mattered most—were okay. But it was an enormous learning experience. For months, they lived in a rental condo in Boulder. They had to literally start over. They had no clothes, no furniture, none of the conveniences that build up in homes over years. The process of dealing with insurance and recovering was more emotional than he expected. He had to inventory what he lost. Somewhere in the midst of that, he says, he was just crushed. It wasn't just the material stuff—a stereo, television, all that. He realized the deep connection he had to some of what was gone—things like his deceased father's watch. He describes an emotional roller coaster. Some days he and his family would laugh. Other days, they were exhausted. They couldn't think about it; they could barely cope.

More than ten years out, his message to others is: "It gets better. It gets easier."

Moraga's story is particularly illuminating because of what this devastating experience meant for his relationship with fire. After losing a home to fire, it's easy to imagine crusading for more suppression, less flame. But Moraga understood how crucial fire is in western landscapes. He had a firm grasp on how flames had been mishandled for a century. "I believe in managing fires," Moraga, the decades-long firefighter, says. "I don't fight fires."

His fascination with fire continued. Eventually, Moraga returned to his home with a different mindset. As a fire behavior analyst, he was "geeking out," he says. He looked at how the fire burnt on different areas of his property, and as the slope changed. It was his own little laboratory. "Fire is not going to change for us, so we've got to figure out a different way to work with it," Moraga says. "If anything, we need more fire. We've got to introduce more fire and we have to find a way to tolerate smoke and live with fire in order to reduce the threat of fire." He started a fire consultancy, focused on assessing and addressing wildfire risk. Moraga says his experience helps him relate to other people living in the WUI. He's been through it. He knows what it means for everything to go wrong. And he also knows the work that needs to be done, and how our relationship with fire needs to change.

SOLASTALGIA

When you see a fire-scorched landscape, is it tragic or…something else?

A sustainable future with fire requires rewiring our relationship with fire, and that requires evolving the ways in which we cope with and think about its impacts too. Our modern era has given rise to a new lexicon required to describe the myriad ways that technology, society, and climate are changing. One word mentioned earlier in this book is *Anthropocene*—an explicit acknowledgment that our current geological epoch is inexorably shaped by humans. Another term crucial to understanding and getting through our predicament is *solastalgia*. Coined in 2003 by the philosopher Glenn Albrecht, solastalgia refers to a sense of grief, loss, or distress as your home changes along with the environment. Think of homesickness or nostalgia for an ecosystem or place that's changed forever.

Climate change is altering environments at an unprecedented pace. The effects of wildfire are a stark example of what might cause solastalgia. A forest burns near a community. Suddenly, community members' backyards are blanketed by ash. Forests where people hunted and hiked are reduced to a pile of matchsticks, often toppled by wind. Streambeds where folks fished and picnicked are choked with mud and silt. Once-lush hillsides are now charred burn scars, constant reminders of fire's destructive effects. Places once familiar are now foreign. For many folks who grew up in the rural West, the spaces that made you "you" changed forever with one wall of flame.

That's, at least, one way to see the "devastation." People tend to get attached to place. We tend to think the way we've known an area is the way it's always been. But one human's existence is a mere blink of an eye in the time span of an ecosystem. The forests we take for granted as "natural," for example, might be overgrown and dense thanks to wildfire suppression. Other effects—of recreation, of cattle grazing, of irrigation, and more—might also have shaped an ecosystem away from its presettlement state.

It's easy for us to forget one of the natural world's most important realities—that ecosystems are never static. As trees grow and die, wildlife populations fluctuate, temperatures warm and cool, the things we take for granted as stable are actually in constant flux. They just change so slowly we don't notice. Wildfire brings that change into sudden relief, often in a way that feels traumatic. Though sudden changes are tough for humans to cope with, they aren't necessarily negative.

Embracing, or at least accepting, that change requires shifting your perspective on the natural world. It isn't easy to do, but there are some practices that could help you come to terms with the ways fires change the landscapes you're deeply attached to.

- Recognize that places change, and there's beauty in that.

- Look over fire-affected landscapes. Walk through them. Listen to what's alive, the sound of water and wind. Sitting and observing helps. Try meditating.

- Find beauty in even the singed and seemingly devastated.

- Notice how quickly plants regrow, especially wild flowers. Appreciate the vivid colors and regrowth. Bring a field guide with you on walks in the woods. Learn about the new growth that's around you.

- Consider morel hunting. These tasty fungi are a sought-after delicacy in areas recently burned. Searching for morels can get you out into the landscape, to find bounty after apparent devastation.

REBUILDING

Across the West, wildfires have leveled towns and entire neighborhoods of cities. The list is long and tragic. Since 2018 alone in California, fires have substantially destroyed Paradise, Greenville, and Big Creek, along with parts of Santa Rosa and Ventura. But the devastation isn't limited to that state. There was Malden in Washington, Talent and Phoenix in Oregon, Carbon in Texas, entire suburban neighborhoods outside Boulder, Colorado. The list goes on and will almost certainly grow in the years to come. These events are emotionally and financially devastating. In the wake of these fires, folks who have lost jobs, homes, pets, loved ones, and places they've spent most of their lives are forced to grapple with what comes next: rebuilding or relocating? Are there some areas just too dangerous to live in?

The data paint an interesting picture of how people rebuild after burns. The research mostly looks at fires from a decade or more ago. One study examined all fires in the U.S. between 2000 and 2005. About a quarter of burnt homes were rebuilt within five years—though the rate was a little higher in the West. However, new development in burnt areas still boomed. Half a decade after fires, there were more homes in burn zones than before the flames hit. One 2021 study looked at the twenty-eight most damaging wildfires in California from 1970 to 2009. The authors found that by and large, there was little adaptation or moving of buildings to lower risk. Ninety-four percent of structures had been rebuilt from thirteen to twenty-five years after the fire, and by and large they were in areas just as risky as ever. People, in essence, didn't learn. Change wasn't happening at the individual level, and local governments were strapped for capacity and reluctant to institute new codes.

It's easy to understand why development sometimes booms after wildfires. Homeowners might not have other options. Their families, their jobs, their entire lives are rooted in place, and few resources are available to help them relocate. Social scientists have studied the connections we develop to place and find that they are multidimensional—they involve family, work, culture, and experience. The disruption of those connections can be devastating. If people do move, where do they go? If it's a large wildfire disaster, the regional housing market might be too small to absorb the folks who have been displaced. So that leaves rebuilding as nearly the only option. At the same time, property prices might be depressed after a disaster, and developers can be eager to swoop in—enticing prospective homeowners who have no experience living in areas prone to burning. After all, the very characteristics that make

WUI towns vulnerable to fire can make them attractive places to live: small, quiet streets, green trees and shady groves, close neighborhoods with quirky houses, life in the mountains.

Most studies on rebuilding, though, focus on fires in WUI communities before the last decade, when fires have caused unprecedented damage. It's worth zooming in on two recent examples to understand more about how communities are rethinking their relationship with wildfire as they rebuild and recover.

Talent, Oregon, and Paradise, California, burnt in 2020 and 2018, respectively. In Talent, a wildfire blazed right through the middle of town, destroying about a third of the homes there. Many were mobile homes. Paradise residents suffered the largest human toll of any wildfire in modern history. Eighty-five people died, and fifty thousand were displaced. Both towns set to rebuilding fast in the years after these tragedies.

By 2022, Talent had already rebuilt more than one hundred of the seven hundred homes lost to the fire. A nonprofit funded by local utility providers gave residents subsidies to build back more efficiently. Some of those upgrades were meant to both make homes more energy efficient—taking small steps against climate change, and therefore wildfires—and make them more resilient to future flames. Locals are using the fire as a way to "reimagine" the city: to make it more walkable, greener, and able to withstand flames whenever they come next. What that means on the ground remains to be seen.

In Paradise, rebuilding began nearly as quickly. By 2021, the California Department of Finance declared the area the fastest-growing city in California. Its population was just over six thousand—still far shy of the more than twenty-five thousand who lived there before the fire. At a press conference, the mayor of Paradise said, "It's essential to make sure we rebuild better, safer and more resilient than before." In June 2022, Paradise became the first town in the nation to require new homes to receive the "Wildfire Prepared Home" designation from the Insurance Institute for Business and Home Safety. The Paradise town council adopted the ordinance unanimously and pursued federal funding to help make it happen. The IBHS's standards focus on many of the mitigation measures we've detailed—from fire-safe roofs to defensible space. As in Talent, the town's recovery plan includes goals to make it both safer and greener.

But the community is also thinking outside the box on how to boost its safety in the future. "You're not going to prevent those catastrophic fires," says Dan Efseaff, district manager for Paradise Recreation and Parks. "We need to learn how to live with fire and adapt to fire and climate." Efseaff helped develop an innovative plan that could create defensible space around the entire community, rather than just individual homes. The idea hasn't been tested at scale, but the notion is relatively simple in theory: create a "greenbelt" or fuel break around the entire area. That would both create limits to urban development as the area grows and provide a space with very little fuel that could slow or stop a fire, as well as offer refuge from the flames, as it approaches. At the same time, the area could provide additional amenities to locals. Paradise residents could use the area to hike, horseback ride, and more.

This doesn't mean something like a moat of cement surrounding town. Rather, it would be a combination of parks, forests, orchards, private properties managed specifically to reduce risk, and more. It's not just a thin ribbon. The area extends from the city limits based on where the city is most at risk. Efseaff says the forested areas would preserve trees, and even utilize prescribed fire. But the woods would look more like they would have a century and a half ago, before fire suppression began. The central challenge is wrangling the patchwork of properties into a sixteen-mile belt to make it happen. Efseaff says in the wake of the 2018 fire, community members view stewardship differently. It's not each person for themselves; they're in it together. If his effort succeeds, it could change the game in Paradise—ecologically and financially. He estimates the buffer could also save local homeowners 30 to 40 percent in insurance premiums, which could make rebuilding there much more affordable.

Efseaff recognizes the buffer zone around the community doesn't fix all risk. A burn the size of the Camp Fire could charge right through it. Plus, megafires can throw embers miles—so homes with wood roofs and other vulnerabilities would still be at risk. But the buffer zone is a start, and an experiment that could feasibly be replicated across the West. So far, Efseaff and his collaborators have assembled over four hundred acres through grants, donations, and arrangements with other local land management entities. He estimates the total effort to complete the buffer will require about $30 million, along with a substantial community buy-in. By June 2022, more than fifteen hundred homes had been rebuilt in Paradise, and hundreds more were under construction. The city expects to reach 75 percent of its population before the fire by 2035.

Locals and officials in both Talent and Paradise—and other towns ravaged by wildfire across the West—aren't following any playbook. But one UC Berkeley report offers policy recommendations for rebuilding after fires: focus on those most vulnerable, meaning low-income people and renters who might be displaced; incentivize low-risk development and punish risky building; build capacity at the city and county level; and to make it all happen, find new funding streams and revenue sources. The report also finds that growth boundaries can help mitigate new development in the WUI and that using tools like conservation easements can help create buffers protecting vulnerable areas. It also emphasizes promoting infill and affordable housing, rather than building more expensive, riskier homes deeper into the forest.

When an entire town is leveled, a blank slate can emerge. In theory, tragedy gives rise to a novel situation: an opportunity to rethink preparedness, safety, climate resilience, equity, and much more from the ground up. Talent and Paradise show that towns might be starting to think differently about their relationship with fire after flames come to their doorstep.

FINDING HOPE IN THE ASH

Fires burning through communities is, without a doubt, tragic. People lose their lives. Families lose their pets and most-prized possessions and memories. Fires can cost taxpayers billions of dollars. Studies consistently find increased rates of PTSD, generalized anxiety, and depression in the wake of fires that affect communities. Those effects linger for years. Emotional injuries can compound on other traumas in people's past too.

There's no minimizing the suffering and sadness that can come with that kind of pain and loss. But there are other lessons too. Disasters destroy, but they also unite. When narratives focus only on the trauma of wildfire—leveled homes, displaced families—it's too easy to lose sight of something else happening beneath the surface.

In her book *A Paradise Built in Hell*, Rebecca Solnit writes about five disasters in North America—from the San Francisco earthquake in 1906 to Hurricane Katrina in 2005. So often, she writes, the media and popular imagination of the aftermath of a disaster is a free-for-all, a dystopian, dog-eat-dog world where every man and woman fend for themselves. But she argues that the reality looks much different and gives us room for optimism. Disasters, she writes, "are a crack in the walls that ordinarily hem us in, and what floods in can be enormously destructive—or creative."

Disasters like wildfire often get painted by the media as times when the worst in humanity emerges: there are often reports of looting, division, and conflict. News stories focus on blame—who set the fire, why did it happen, what mistakes did the Forest Service make? But Solnit's analysis says that's getting so much wrong. "The prevalent human nature in disaster is resilient, resourceful, generous, empathetic, and brave," she says. Disasters, in this sense, show us who we really are.

Solnit's book doesn't focus on wildfire. But her thesis applies just the same. Disaster, she argues, "drags us into emergencies that require we act, and act altruistically, bravely, and with initiative in order to survive or save the neighbors, no matter how we vote or what we do for a living." In the aftermath of the Camp Fire, the Marshall Fire, any fire that's torn through a community, a similar picture forms in the wake of the flames. Neighbors help neighbors find shelter and food. They offer comfort and solace. There are massive drives for food and basic supplies to help firefighters and the displaced. Makeshift kitchens appear. Local groups have formed on the spot to assist with cleanup, resilience, and whatever comes next. Yes, there's trauma, but there's also kindness, altruism, bravery, empathy, and compassion.

This book is meant to help people across the country stay safe from wildfire. Doing that requires accepting a lot more wildfire—and not always putting it out. It also means stepping up to your role in the problem and its solution. Though this starts on your own property, to get where we need to be, communities across the West need to mobilize together. Neighbors who've never met need to come together and make the effort to understand one another. People need to talk and share and help one another. Divides need to be bridged. In this way, addressing wildfire also presents the opportunity to take meaningful action on other major social problems, especially climate change. The window after disasters like wildfire shows us that, even in times as isolating and divisive as they are now, the sort of unity we need already exists within each of us.

The lesson of disasters like wildfire isn't that the government will swoop in and save you; it's that we must save one another. The meaningful, immediate help is never top-down. It comes from other community members. That is where hope lies: in one another. It lies in working together, and in connecting deeply and with empathy. Fire can create what literally looks like an ashen hellscape. But shifting focus—just the same as one might do to see the beneficial effects of wildfire on the landscape—shows what community members are capable of. It shows that we can change. We can work together. And we can tackle the enormous challenge ahead.

CONCLUSION

When you imagine fire season in the West ten, twenty, even thirty years from now—what does it look like?

A pessimist might say: greenhouse gas emissions have continued to rise, causing global temperatures to skyrocket. Drought, heat, and fire are hallmark characteristics of the landscape. The country's firefighting apparatus is larger than ever, hell-bent on snuffing out any and all flames. But still, fire seasons continue to routinely break records in terms of acreage, property damage, and lives lost. Firefighters can't keep up. Against the now common megafires, they don't stand a chance. Entire communities and beloved forests have been charred to ash. The millions of acres of matchstick trees feel downright dystopian. For months of the year, it seems, skies are thick with gray and orange smoke. To avoid that toxic, ashy air, you and your neighbors further retreat to indoor spaces and your screens, more disconnected than ever. Trust in one another and in government—already perilously low today—collapses. There's an exodus from the West, where communities lie in ruins.

In this version of our future, the wildfire problem is overwhelming and insurmountable, along with climate change. If we keep on doing what we've been doing, it's not all that difficult to imagine. But the thing is, it doesn't have to be this way.

Here's another version of our fiery future: wildfire is a defining feature of the West. But the flames we see aren't nearly as destructive. Low-intensity burns flourish when the time is right. Prescribed fire, led by Indigenous groups and local communities—not just the feds and state foresters—is the norm in the spring and fall. The way fires are fought has shifted too. Instead of quashing all fires, many are carefully given room to burn as they naturally would.

The scars of those fires create barriers to the spread of extreme burns and help restore ecosystems so that native flora and fauna flourish. Government agencies like the Forest Service productively collaborate with tribes, ranchers, and community members. Those stakeholders lock eyes, share ideas, and learn from one another.

In this version of our future, there's been an ideological shift. Together, we recognize that finding a healthy relationship with fire requires letting go of the idea that we—humans—are in charge of the natural world. It also demands recognizing the myth that we can control and bend fire to our will for what it is: a fiction. There's broad tolerance for occasional smoke in the skies. And smoke does, indeed, float in at certain times of the year. But it isn't a suffocating fog that sits for months on end. It's short-lived, and we're ready for it. It's easy to plan for and coexist with.

At the same time, metal roofs have sprung up in wild-fire-prone areas, like morels after a burn. Across the West, neighbors have mobilized to help entire communities make their properties more fire resilient. That includes work as mundane as raking up yards and replacing vents on the outside of homes or reconfiguring the entire exterior shells of houses and the "defensible space" within one hundred feet of their walls. Folks are taking advantage of the billions of federal and state dollars available to help that work. Counties and cities have enacted zoning laws and other regulations that limit development in areas that are particularly vulnerable to flames.

We've changed our interactions with one another, as well. Instead of isolating people in terror, wildfire has inspired and united. Community members in the WUI get together routinely, in person, to talk about their visions of where they live and how to achieve them. Neighbors help neighbors spread the word about evacuations when necessary, and help one another learn what work is crucial to fireproof their homes, and how to get it done. The lessons we learn preparing for wildfire help us learn how to live better with one another. Those lessons help reconfigure our relationship with the natural world that we're a part of.

We finally seem capable of long-term thinking. This action on wildfire is coupled with action on climate change, as well. Many of those metal roofs have solar panels on them. The electric grid is nearly entirely decarbonized. Green jobs have sprung up from coast to coast, bringing life and livelihood to industrial towns—including areas once reliant on timber. Nationally, the country has reached a milestone known as "net zero emissions," meaning fewer emissions are produced than are removed from the atmosphere every year. As the country eases the pedal on climate-warming greenhouse gases, it's charting a course to less severe fire seasons. This pays off not just with fire but in politics, in climate, in all the things that matter for making a better future for our children and grandchildren.

One of these two futures could very well be our reality. The choice is ours to make.

One way to conceptualize a sustainable existence with flames draws from the ideas of the legendary conservationist (and U.S. forester) Aldo Leopold. His "land ethic" calls for reconfiguring our moral relationship with the natural world. In his seminal book, *A Sand County Almanac*, he writes, "A thing is right when it tends to preserve the integrity, stability, and beauty of the biotic community. It is wrong when it tends otherwise." Leopold also encouraged "thinking like a mountain." That means holistic thinking, considering all parts of the natural world, including us, as interrelated parts of a system. A mountain, after all, doesn't plan in terms of the week ahead or even election cycles. A mountain forms over thousands of years and exists at the nexus of innumerable ecological forces (including, of course, wildfire). It's easy to think only in terms of ourselves and our immediate surroundings, in the moment. But thinking like a mountain requires recognizing the interconnection of people, place, and ecosystem.

Think about these ideas, and how they relate to fire. There could be no natural world, no Leopoldian "integrity, stability, and beauty" of biotic communities in forests across the West—even, across the globe—absent fire. But it also doesn't mean that all fire is good. Instead, a land ethic might help us appreciate the kind of fire we need: the low-intensity, periodic burns required to keep our ecosystem in balance. A high-intensity megafire that scorches soil to near death certainly doesn't do any favors for the stability of the biotic system. Leopold's ideas aren't perfect. But they're a useful thought experiment, especially when it comes to wildfire. His land ethic tells us that it's not only a good idea ecologically but a moral responsibility to reconfigure our relationship with fire—and also with other human activities that throw natural systems out of whack, like burning fossil fuels.

This book has few of the stories that typically dominate wild-fire media—narratives like the heroism of firefighters narrowly escaping a burn, or the loss and despair and terror as a wildfire approaches a neighborhood. That's intentional. While these are realities of wildfire—and they're often tragic—they're also in many ways outliers. They crowd out other narratives that deserve even more attention. Fire isn't always bad. It isn't an enemy we must fight and eradicate. It's part of our planet. Fire is guaranteed in our future, and forging a sustainable path ahead requires learning how to live with it. That isn't necessarily easy to do, but the solutions exist now.

We can protect our homes, protect our communities, and protect the landscapes we love. To make it happen, we all need to do three things. First, "harden" our homes and the one hundred feet surrounding them so that they are more likely to withstand fire. Second, tolerate—even advocate for—*more* fire on the landscape, not less. Third, talk to one another. Help your neighbors think long-term, understand how we got ourselves into this fiery predicament, and how we can help one another get out of it. At the same time, state and federal governments need to fundamentally change how they think of fire too. Enormous investments are needed to help all people in the West—not just the wealthiest—improve the fire resilience of their homes and properties (and therefore families). Counties and localities need to make tough decisions about zoning and regulations. Firefighting agencies need to improve collaboration, communication, and transparency, and cultivate broad tolerance for ecologically appropriate natural fire where it's safe to roam free. Our elected officials must stop demonizing wildfire and ramp up policies that lower emissions and address the worst impacts of climate change.

All that will take individual vigilance, community organizing, policy change, and an immense reallocation of resources. And this requires a change so much deeper than just policy. "We just don't have a culture of having fire; we have a culture of *not* having fire," Mark Finney, a Forest Service research forester, says. "So when it comes to trying to change that, you're overturning generations of belief." Changing beliefs means changing culture. It means changing ideology.

Overturning generations of belief—a whole ideology—is a daunting task. But there's an optimism in the work that lies ahead. Lenya Quinn-Davidson calls the small networks of prescribed burning associations popping up across the West a "social movement." That's because it *is*. Those burners are thinking of fire in new ways. And those groups, along with tribes across the country, scientists, even Forest Service employees, are paving a new path for fire. Combined, there's a groundswell forming. The change needed to shift culture will always come from the ground up, not the top down. It will come from folks like you, from communities like yours.

In this way, doing the work that's crucial to saving your own home and community is a revolutionary act. Getting a new roof, installing one-eighth-inch eave vents, and pruning and cutting trees and shrubs to create defensible space are downright radical. So is calling for more fire on the landscape near your home, rather than less. And, of course, organizing your neighbors. That's because learning to live with fire requires learning to live better with one another. And that pays off. It plants the first planks of a bridge to a better world for so much more than just wildfire.

Living better with one another is especially important because the generations of belief that need overturning aren't just related to wildfire. Lessening the severity of wildfires also requires reducing our emissions. That, like changing our relationship with flame, requires shifts in both policy and collective action, and in individual values. In this way, changing our relationship with wildfire is part of a much bigger project: changing our relationship with the world around us.

The work ahead will not be easy. But we can do it. We have to.

RESOURCES

InciWeb offers information on all the fires currently burning across the country. Visit the website for real-time updates on wildfires near you. Those updates include maps, evacuation notices, briefings, fire size, resources devoted to suppression, and more. Go to inciweb.wildfire.gov.

AirNow is a centralized repository of air quality information. You can search by your zip code and find the wildfire smoke information you need. Learn more about what wildfires mean for the air you're breathing at airnow.gov.

National Fire News, by the National Interagency Fire Center, offers summaries of fire activity across the country. Updates also include the resources ready to tackle fires as they emerge and weather patterns that may affect fire behavior. Navigate the rest of NIFC's website for fire statistics over the years and lots of other helpful information at https://www.nifc.gov/fire-information/nfn.

Firewise USA offers a set of standards and practices that can help make your home more resilient to wildfire. Its website also provides resources about communities already taking part in their program, success stories, and more. Learn more at https://www.nfpa.org/Public-Education/Fire-causes-and-risks/Wildfire/Firewise-USA.

As the climate changes, so do the risks to your home. **Risk Factor** offers projections of wildfire risk to your home, going out to thirty years. It also has information about flood and heat risks. Search by your home address and get a good picture of what's coming your way. Learn more about your risk at riskfactor.com.

The **Community Wildfire Planning Center** offers a helpful guide that breaks down wildfire information and resources by state. Find out about your area at https://www.communitywildfire.org/resources-by-state/.

Creating defensible space around your house is crucial to making sure that your home can survive a wildfire should one burn nearby. There are lots of resources online to help with this, but **Cal Fire**'s guide is particularly helpful: https://www.readyforwildfire.org/prepare-for-wildfire/get-ready/defensible-space/.

Headwaters Economics offers a plethora of information about home preparedness in the wildland urban interface. The **Community Planning Assistance for Wildfires** portal is particularly useful for those interested in helping entire communities prepare for wildfires. Go to https://cpaw.headwaterseconomics.org/.

Fire Adapted Communities Learning Network is a resource for anyone living in a community vulnerable to wildfire. It offers workshops, lessons, and much more. Peruse the site for creative ideas about fireproofing your home and community, and to connect with other individuals and communities also grappling with resilience: https://fireadapted.org/.

So much wildfire preparedness occurs at the local level. In addition to using the above resources, be sure to look into actions and resources in the state, county, and city where you live.

ACKNOWLEDGMENTS

This book would not have been possible without, first, a podcast. The show *Fireline* was made possible with the generous support of Montana Public Radio and the University of Montana College of Business. We also couldn't have created the show—or continued our deep thought about living with wildfire—without producer Victor Yvellez's exceptional work.

Jessy Stevenson's wonderful artwork illuminates key concepts throughout this book. Her beautiful illustrations deliver important nuance that goes beyond the meaning our words can convey. Our editor and publisher, Anton Mueller, believed we could write this book, perhaps before we did. Morgan Jones and her Bloomsbury colleagues took our words and made them something people could hold. And our wonderful agent, Danielle Svetcov, guided us with clarity and care throughout the process.

We must also thank every single person who talked with us for both the podcast and this book, which are too many to name in this short section. In particular, we want to thank the late Tony Incashola Sr., and his son, Tony Incashola Jr., for having faith in us as storytellers and listeners and showing us their controlled burn on tribal land.

There are also folks Nick and Justin would like to thank individually. Nick couldn't have written any of this without the support of Leah Swartz, without whose understanding, thought, and patience he could never have taken the steps to create this book. Michael Kodas showed him what great journalism about wildfire could look like. Amy Martin taught him basically everything he knows about storytelling. Justin is deeply grateful for the support of his wonderful family—Maggie, Ainslie, and Charlotte. Wayne Williams, Brent Ruby, and Dan Cottrell opened a doorway into the wildland firefighting community and were trusting enough to share their stories. Eric Legvold supported *Fireline* and this book from the get-go, especially when things got hard.

We'd both also like to thank all the folks working to change how we live with wildfire. That includes firefighters and fire managers who treat fire as an ally, not an enemy, and work to change the way we engage with it. It also includes people in communities working to bolster resilience, to collaborate with one another on preparedness practices, and to prepare ahead of time for the next burn. More than anything, it includes the people hell-bent on bringing more fire back to the land, not less. There's extraordinary work happening on all these fronts. The stories in this book are just the beginning.

A NOTE ON THE AUTHORS

NICK MOTT is a journalist and podcast producer. His podcast work has received a Peabody and two National Edward R. Murrow Awards. His print and audio reporting has been published in the *Atlantic*, NPR, *High Country News*, and the *Washington Post*, among many other outlets.

JUSTIN ANGLE is a professor and the Poe Family Distinguished Faculty Fellow at the University of Montana College of Business. His work has been published in *Journal of Marketing*, *Journal of Consumer Research*, and the *Washington Post*.

Together, Justin and Nick created *Fireline*, a national Edward R. Murrow Award–winning podcast about wildfire in the West. They live in Montana.